PLANT SCIENCE RESEARCH AND PRACTICES

JATROPHA CURCAS

BIOLOGY, CULTIVATION AND POTENTIAL USES

Plant Science Research and Practices

PLANT SCIENCE RESEARCH AND PRACTICES

JATROPHA CURCAS

BIOLOGY, CULTIVATION AND POTENTIAL USES

GREGORY MEDINA
EDITOR

nova publishers
New York

Additional color graphics may be available in the e-book version of this book.

Library of Congress Cataloging-in-Publication Data

Jatropha curcas : biology, cultivation and potential uses / editor: Gregory Medina.
pages cm. -- (Plant science research and practices)
Includes index.
ISBN 978-1-63483-089-8 (hardcover)
1. Jatropha. 2. Jatropha--Utilization. I. Medina, Gregory, editor. II. Series: Plant science research and practices.
QK495.E9J37 2015
583'.69--dc23

2015019754

Published by Nova Science Publishers, Inc. † New York

CONTENTS

PREFACE

Jatropha curcas Linnaeus is a multipurpose plant belonging to the Euphorbiaceae family. It has social, agricultural, environmental, industrial, pharmaceutical and energy production potentials. It has high amounts of oil in its seeds, which can be converted to biodiesel. This book begins with a discussion on the use of *Jatropha curcas* as a biofuel. It continues by exploring other potential uses of *Jatropha curcas.*

Chapter 1 – Biofuels have attracted considerable attention over the last 20 years as a renewable alternative to fossil fuels for energy production. Biofuel use is also increasing rapidly all over the world and it is expected to have an increasing impact on both energy and agricultural sectors. Different studies have been carried out using different vegetal material to produce biofuels, like *Jatropha curcas*. This is a multipurpose plant that contain high amounts of oil in its seeds, which can be converted to biodiesel. In addition, the cellulose obtained from its oilseed press cakes (a byproduct from biodiesel production) can be converted into ethanol through sequential steps of acid pretreatment, hydrolysis, and fermentation by the yeast *Saccharomyces cerevisiae*. *Jatropha* fruit is also available in significant amounts on annual basis. In addition to the high amount of oils found in their seeds, species of *Jatropha* are characterized by having edible or inedible fruits with high levels of lignocellulosic material for conversion to fermentable sugars. Biogas production is also possible from seed cake and husk capsules by anaerobic fermentation. In this chapter the current knowledge on the *J. curcas* plant uses is reviewed, mainly in the biofuel production.

Chapter 2 – Advances in sequencing of genomes has lead to the promise of genetically modified fuel crop such as Jatropha Curcas that is used to obtain Jatropha Oil. The gene regulation via Ribosome inhibiting Protein in the shrub has been studied using transfection of clones of curcin genes found in Jatropha Curcus and in other plants. This can lead to better cultivability of the shrub and higher yield of Jatropha Oil. NGS, Next-Generation Sequencing machines available commercially are tabulated. The triglycerides in Jatropha Oil can be transesterified using excess methanol in the presence of catalyst or excess methanol into biodiesel and glycerol. The glycerol can be sold as by-product. The kinetics of consecutive-competetive reactions during methanolysis of triglycerides with intermediates such as diglyceride, monoglyceride are modeled. The model solutions assuming irreversible simple first order kinetics were obtained using the method of Laplace transforms. Although biodiesel forms in each consecutive step under some conditions of reaction rate ratios α, κ the yield of glycerol are seen to be higher than that of biodiesel. This can be called "cross-over" and can be seen in the illustrations shown as prototypical examples. The operating cost of separating biodiesel from glycerol using centrifuge can be obtained from computer simulations of layer formation during shear flow. For optimal total cost there exists a yield of biodiesel where the process can be operated at.

Chapter 3 – The member of eupherbiaceae family, Jatropha is being explored as multi dimensional prospective plant. It has social, agricultural, environmental, industrial, pharmaceutical and energy production potentials. Jatropha seed cake, compost from debris and slurry from biogas plant are good sources of soil nutrition. Protein from seed cake can be utilized as animal feed. As a plant Jatropha can assimilate CO_2 from climate and it can also be used as bio adsorbent of heavy metal from effluents. Protein, lignin and other products of Jatropha can substitute the hazard synthetic chemicals of polymers production. Seed oil have been using in soap, dying and cosmetic industry from long before. Extract from Jatropha leaf, bark posses biological active compounds that have medicinal potentials. Jatropha is more familiar as energy crop. It provides energy in the form of firewood, seed oil as lamp, cooking and engine fuel directly. Biodiesel the chemical conversion of seed oil can be used directly or blend with fossil diesel with modified or conventional engine. However, presence of toxic substances (pherbol ester) renders its utilization as therapeutic agents and animal feed. High viscosity of seed oil and biodiesel is the hurdle to use. More importantly, poor yield in field condition than expectation is the main barrier to be commercialized of Jatropha cultivation. It needs to extensive and systematic screening of available germplasoms and application of biotechnology for the development of high yielding cultivar, agronomic practice and crop management for maximum yield, biochemical engineering for the improvement of biodiesel quality and finally customization of engine according to fuel properties.

Chapter 4 – *Jatropha curcas* L. is a popular multipurpose crop that is extensively grown in many tropical countries. Its popularity has been ascribed to its acclaimed attributes as a non-food crop that grows well in marginal land not suitable for crop production, relatively low requirements of water, fertilizer and maintenance. It is arguably true that Jatropha can provide additional income in areas where there is plenty of land and labour is readily available. As such, Jatropha is now a component of farming systems in many developing countries. Traditionally, farming systems in developing countries have focused on the twin purposes of food and fibre production. The entrance of Jatropha adds a third dimension to this model. This arises mainly from the multiple uses of Jatropha, particularly energy production. The possibilities of exploitation of Jatropha for various uses have been explored and continue to be explored. This has largely been underpinned by technological innovations for processing different parts of Jatropha into value-added products. The major parts of Jatropha that can be exploited for multiple uses include the woody parts, seed, shell, husks, kernel, oil and press-cake. Press-cake is a by-product of oil extraction. Mature technologies exist for converting Jatropha components into useful products. The major focus has been on developing appropriate technologies for converting Jatropha components into first and second generation energy careers. The gamut of technologies available for conversion purposes ranges from biological to chemical methods. Uses in soap and fertiliser production, as food, medicine and pesticide have been explored. Interfacing the gamut of technologies available and the multiplicity of uses of Jatropha, provides an opportunity for promoting what can be called the Jatropha entrepreneurship system. There is no doubt that Jatropha can be more profitable promoting small to large-scale entrepreneurship if production is coupled with appropriate management practices and seed oil is processed into other valuable products other than biofuel. Entrepreneurship is a major driver of economic development for large populations in developing countries. As such, crops like Jatropha provide opportunities through the various products that can be obtained from them. In this chapter the authors develop a model for promoting entrepreneurship based on the multiple uses of Jatropha and use of available technologies, emphasising the benefits to rural economies.

Chapter 5 – In December 2007, seeds of *Jatropha curcas* from two African and six American provenances were introduced in Tunisia for the first time. A plantation was established in Southern Tunisia (region of Gabes). All plants were irrigated with treated wastewater. In 2013 and 2014, height, canopy circumference, seed yield and percentage of aborted seeds were studied for six and seven-year-old *J. curcas* plants. In 2013, four fructification periods were registered.

The highest seed yield was observed for a Brazilian provenance (A4) and its average was 398 g/plant. The lower was for A8 (Suriname provenance) with 134 g/plant. In 2014, six fructification periods were observed, the highest height and seed yield were registered for A3 (Brazilian provenance); their values were 297.5 cm and 804 g seeds/plant/year, respectively, and the lower seed production was reported for a Brazilian provenance (A2; 317 g/plant). In winter, seed yield is high, but it is characterized by a high percentage of aborted seeds. Oil content of seeds harvested in August 2013 varied from 23.46% (A7) to 32.15% (A5).

Chapter 6 – *Jatropha carcus* is an important plant for the commencement of energy plantation. Jatropha can be grown in marginal, degraded, waste lands in tropical and subtropical regions of the world. It is a partially domesticated plant hence development of high yielding cultivars and package of management practices needs prime importance. The most important use propagation is to produce high quality planting materials. Jatropha can propagate through seeds, cuttings and tissue culture. Due to poor seed germination, short viability and high heterozygosity seed propagation may not produce high quality planting materials. On the other hand propagation through vegetative means or micropropagation can be the most effective alternatives to produce quality seedlings with desirable traits for higher seed and oil yield. Vegetative propagation has inimitable significance to maintain homogeneity among the progeny and provide early production.

The aspects of various propagation methods used in Jatropha cultivation such as seed propagation, vegetative propagation and micropropagation are highlighted in this chapter.

Chapter 7 – The study concerns the effect of the aqueous extract of physic nut (*Jatropha curcas* L.) on the germination of six cultivated crops: *Hordeum vulgare* L., *Medicago sativa* L., *Corchorus olitorius* L., *Trigonella foenum-graecum* L., *Lens culinaris* L. and *Cicer arietinum* L. The extracts were obtained after an incubation of the dry plant material in distilled water (9%) during 48 hours at 60°C. The extraction was made from five organs of *J. curcas* (root, twig, leaf, seed and pericarp). The results showed a variable behavior between studied organs and species. *Corchorus olitorius* is the most tolerant species, showing elevated germination rate (> 90%) for all treatments. *L. culinaris* is the most sensitive species; its rate of germination is lower than 23% for all treatments with *J. curcas* aqueous extracts. The pericarp aqueous extract is the most inhibitive of the germination; the authors obtained rates of 0% for *T. foenum-graecum, L. culinaris* and *M. sativa*. The aqueous extracts of *J. curcas* affect also root length; a reduction of the parameter was observed for all studied species, essentially with the aqueous extract of pericarp. Allelochemicals may be present in any parts of *J. curcas* but fruit pericarps are the major source.

Chapter 8 – The removal of phorbol esters (PEs) from *Jatropha curcas* seed oil by adsorption onto silica gel was investigated. The average PE content in the oil, as determined by HPLC, was 1.68 mg TPA/g oil. The optimized conditions for PE adsorption by silica gel were studied. The results showed that the highest PEs removal (77.5%) and PE adsorption (104.2%) were obtained when 0.3 g of silica gel/5 g of seed oil was used at ambient temperature (30 °C) for 60 min of adsorption time and with an agitation rate of 250 rpm. The characteristics of the silica gel, as determined by FT-IR, indicated that PEs were adsorbed onto the surface of the silica gel. A comparison of the physico-chemical properties of the raw oil and the oil obtained after adsorption demonstrates that the adsorption by silica gel let to receive lightens the oil color without changing its fatty acid composition or introducing contamination from other compounds. Additionally, the results showed that silica gel can be used two additional times in subsequent PE adsorptions without a significant loss in reusability. In conclusion, adsorption by silica gel can remove PEs from *J. curcas* seed oil. This method offers simplicity of design, operation and scale-up, avoids using toxic solvents and minimizes the degradation of PEs.

Chapter 9 – This study aims to control the combined effect of water stress and nitrogen deficiency on growth and physiology of plants of Jatropha curcas, aged one year and issued from two different origins provenances: PAR (Paraná-Brazil), SUR (Surinam- North South America). The water deficit was applied during the month of October 2011. Two water regimes were adopted (I: irrigated plants, NI: non irrigated plants) associated with two nitrogen regimes (N-: no nitrogen, N+:2 g of nitrogen). Some physiological parameters were measured (leaf chlorophyll content, soil moisture and relative water content on leaf) as well as growth parameters (Stem diameter of the stem and leaf area, leaf thickness) were monitored during this survey. The results revealed that the lack of water and nitrogen decrease the averages of stem diameter and the leaf area: This was more emphasized on Surinam provenance. The authors noticed also an increase on leaf thickness especially on Paraná provenance. Furthermore, the ratio root biomass / aboveground biomass had increased more with Paraná provenance; this ratio indicated best tolerance to drought. Chlorophyll content in the leaf and water content in the soil are very sensitive to the combined deficit of water and nitrogen: These were significantly reduced on Surinam provenance. However, the relative water content of leaves were not often affected by water deficit especially on Paraná provenance, and were generally maintained at relatively high values (>50%).

Chapter 10 – Mexico is a country with a great variety of natural resources like seeds and grains that can be employed to improve bread manufacture. Recent investigations shown that *Jatropha* flour can be used to improve bread nutritional properties. The aim of this work was to obtain optimal mixtures of wheat flour, barley concentrate and *Jatropha curcas* flour, to develop a bread product (white bread), evaluating doughs and breads mechanical properties. *J. curcas* and barley concentrate doughs mechanical evaluation indicated that the optimum formulation to bread formulation was with 3.7% of both ingredients. The hardness, adhesiveness, cohesiveness and elasticity of this bread were lower than the control bread. Control bread crumb quality presented no-homogeneous and more open alveolus, as compared with the closer and more uniform crumb in the formulated bread, with lower loaf volume. Higher fiber content in barley concentrate resulted in higher bread weight, with higher water retention capacity. Proximal chemical analysis showed a protein content of 11.05 and 7.47 of fiber.

In: Jatropha Curcas
Editor: Gregory Medina

ISBN: 978-1-63483-089-8

Chapter 1

JATROPHA CURCAS AS A BIOFUEL

Eugenio Sánchez-Arreola[1*], Luis R. Hernández[1] and Horacio Bach[2]

[1]Departamento de Ciencias Químico Biológicas,
Universidad de las Américas Puebla, Cholula, Puebla, México
[2]Department of Medicine, Division of Infectious Diseases,
University of British Columbia, Vancouver, Canada

ABSTRACT

Biofuels have attracted considerable attention over the last 20 years as a renewable alternative to fossil fuels for energy production. Biofuel use is also increasing rapidly all over the world and it is expected to have an increasing impact on both energy and agricultural sectors. Different studies have been carried out using different vegetal material to produce biofuels, like *Jatropha curcas*. This is a multipurpose plant that contain high amounts of oil in its seeds, which can be converted to biodiesel. In addition, the cellulose obtained from its oilseed press cakes (a byproduct from biodiesel production) can be converted into ethanol through sequential steps of acid pretreatment, hydrolysis, and fermentation by the yeast *Saccharomyces cerevisiae*. *Jatropha* fruit is also available in significant amounts on annual basis. In addition to the high amount of oils found in their seeds, species of *Jatropha* are characterized by having edible or inedible fruits with high levels of lignocellulosic material for conversion to fermentable sugars. Biogas production is also possible from seed cake and husk capsules by anaerobic fermentation. In this chapter the current knowledge on the *J. curcas* plant uses is reviewed, mainly in the biofuel production.

* Corresponding author. Tel.: +52 222 2292410; fax: +52 222 2292419. E-mail address eugenio.sanchez@udlap.mx (E. Sánchez-Arreola).

INTRODUCTION

Biofuels have attracted considerable attention over the last 20 years as a renewable alternative to fossil fuels for energy production. Biofuel use is also increasing rapidly all over the world and it is expected to have an increasing impact on both energy and agricultural sectors. Biofuel production from agricultural crops is a potential alternative to fossil-based gasoline or diesel, and some crops like *Jatropha curcas* have been utilized for this purpose due to its high content of oil that can be used as raw material for fuel production.

The availability and sustainability of sufficient supplies of less expensive feedstock in the form of vegetable oils from plants like *J. curcas*, and the establishment of efficient processing technology to produce biofuel, is critical to production, implementation, and use of competitive biofuels, especially biodiesel. Numerous studies have reported the oil content, physicochemical properties, and fatty acid composition of *J. curcas*. Taking into account all these previous reports, the fuel properties of *Jatropha* biodiesel is comparable to fossil diesel.

J. curcas Linnaeus is a multipurpose plant containing high amounts of oil in its seeds, which can be converted to biodiesel. In addition, the cellulose obtained from its oilseed press cakes (a byproduct from biodiesel production) can be converted into ethanol through sequential steps of acid pretreatment, hydrolysis, and fermentation by the yeast *Saccharomyces cerevisiae. Jatropha* fruit is also available in significant amounts on annual basis. In addition to the high amount of oils found in their seeds, species of *Jatropha* are characterized by having edible or inedible fruits with high levels of lignocellulosic material for conversion to fermentable sugars. Biogas production is also possible from seed cake and husk capsules by anaerobic fermentation.

The objective of this chapter is to summarize current literature on the *J. curcas* plant, including:

- Production of biofuel from seed oil and lignocellulosic material.
- Production of methane (by anaerobic digestion) and bioethanol.
- Fuel properties of the biodiesel and attempts to improve vegetable oil to biodiesel conversion.
- Advantages and disadvantages of biofuel technologies used to prepare biodiesel, bioethanol and biogas, including biotechnological solutions.
- Life cycle assessment and environmental impact of large-scale plantations.

1. PRODUCTION OF BIODIESEL FROM THE SEED OIL

1.1. Extraction of *Jatropha* Oil

J. curcas Linnaeus (JC) is a multipurpose plant belonging to the Euphorbiaceae family. The plant produces flowers six months after sowing and can generate 0.4 to 10 kg of seeds per plant annually depending on the age of plant, soil, and climatic conditions [1]. The estimated potential area of JC plantations is 59-1486 million hectares worldwide, with potential production of 56-3613 million tons of dry seed yearly [2].

The high oil content of JC seed (40 to 63%) indicates that they are suitable as inedible vegetable oil biodiesel feedstock [3, 4]. Different methods have been used to extract the oil, such as extraction by soxhlet apparatus [3, 4], seed pressing [5], supercritical carbon dioxide [6, 7], ultrasonic-assisted extraction [8, 9], microwave-assisted extraction [10], and aqueous enzymatic oil [9]. Sayyar et al. (2009) investigated the effects of five operating parameters on oil extraction by developing a model that included solvent type, temperature, solvent to solid ratio, processing time and particle size of the meal, to optimize the processing conditions for achieving maximum oil yield from JC seed. The kinetics of extraction was assumed based on a second order mechanism. The initial extraction rate, the saturated extraction capacity, the rate constant of extraction and the activation energy were calculated using the published model. The optimum conditions were found to be an 8 h reaction time, temperature of approximately 68°C, coarse particle size (0.5-0.75 mm), solvent (hexane) to solid ratio of 6:1, and the activation energy was 8021.9 J moL^{-1} [11].

1.2. Physiochemical Properties of Oil

The quality of a specific oil is determined by its physicochemical properties, such as acid value, cetane number, density, flash point, heating value, iodine value, pour point, refractive index, saponification values, specific gravity, and viscosity. Briefly, the acid value is a measurement of the extent to which esteric bonds of fatty acids are hydrolyzed into the parent glyceride molecule, and it is expressed as the content of Free Fat Acids (FFA) in the oil. The cetane number measures the ignition performance of a fuel by comparing it to reference fuels in a standardized engine test, such as the American Society for Testing and Materials (ASTM) 613. It is generally dependent on the composition of the fuel and can impact the engine's startability, noise level, and exhaust emissions. The flash point temperature measures the tendency of the tested material to form a flammable mixture with air under controlled laboratory conditions. This number must be considered when assessing the overall flammability hazard of a material (ASTM 63). The iodine value is a parameter that indicates the degree of unsaturation and the trend of the oil to oxidize.

Table 1. Physiochemical Properties of JC Seed Oil

Parameter	Values	References
Acid value (mg KOH/g)	0.7-28.5	[3, 12-16]
Cetane number	35.37-57.00	[12, 13]
Density (20°C)	0.902-0.920	[3, 4, 16]
Flash point (°C)	125-225	[13, 16]
Caloric value (MJ/kg)	37.07-40.30	[12, 16]
Iodine value	83.5-187.3	[3, 4, 14, 15]
Pour point (°C)	-2 to 1	[12, 13]
Refractive index (20°C)	1.473	[15]
Saponification values (mg KOH/g)	183.3-211.9	[3, 4, 14, 15]
Specific gravity (20°C, g/cc)	0.904-0.920	[3, 12, 15]
Viscosity (40°C, mm^2/s)	8.72-34.84	[12, 13, 17]

The physiochemical properties of JC seed oil vary according to the climate and the location where the plant is cultured. The available data of reported physiochemical properties of JC oils is summarized in Table 1.

The fatty acid composition of JC seed oil of different origins is compared in Table 2. *Jatropha* seed oil has 79% unsaturated fatty acids on average, with oleic acid predominant followed by lenoleic acid, and 21% saturated fatty acids with palmitic acid predominant followed by stearic acid.

Table 2. Fatty acid composition from different varieties of *Jatropha curcas* seed oil (%)

Fatty Acid	Malaysia [4]*	China [6]*	Africa [14]*	Nicaragua [14]*	India [18]*	Mexico^ [15]*
Capric 10:0	NR	NR	0.1	0.1	NR	0.9
Myristic 14:0	0.1	NR	0.1	0.1	NR	0.4
Palmitoleic 16:1	0.7	0.3	0.9	0.8	NR	0.5
Palmitic 16:0	14.2	11.7	15.1	13.6	12.8	16.1
Margaric 17:0	0.1	NR	NR	NR	NR	NR
Linolenic 18:3	0.2	NR	0.2	0.2	NR	NR
Linoleic 18:2	32.8	18.5	31.4	43.1	34.0	27.6
Oleic 18:1	44.7	52.7	44.7	34.4	44.8	47.2
Stearic 18:0	7.0	4.4	7.1	7.4	7.3	7.0
13-eicosenoic 20:1	NR	2.8	NR	NR	NR	NR
Arachidic 20:0	0.2	0.3	0.2	0.3	NR	NR
13-docosenoic 22:1	NR	9.2	NR	NR	NR	NR
Behenic 22:0	NR	NR	0.2	NR	NR	NR
Others	NR	0.1	NR	NR	1.1	0.3
Σ Saturated fatty acid	21.6	16.4	22.8	21.5	20.1	24.4
Σ Unsaturated fatty acid	78.4	83.5	77.2	78.5	78.8	75.3

* Reference number. ^ Edible plant. NR, not reported.

1.3. Transesterification

Biodiesel, constituted by fatty acid alkyl esters (FAAE), is obtained by transesterification or alcoholysis, defined as the transformation of a fat or oil with an alcohol to form FAAE and glycerol (Figure 1). The substrates for the reaction can be common vegetable oils or animal fats, which are esters of saturated and unsaturated monocarboxylic acids with trihydric alcohol glyceride. These esters are called triglycerides, which can react with alcohol in the presence of a catalyst. During the stepwise conversion of triglyceride into diglyceride, monoglyceride, and finally to glycerol, 1 mol of fatty ester is liberated at each step. Usually, methanol is the preferred alcohol for producing biodiesel because of its low cost [19].

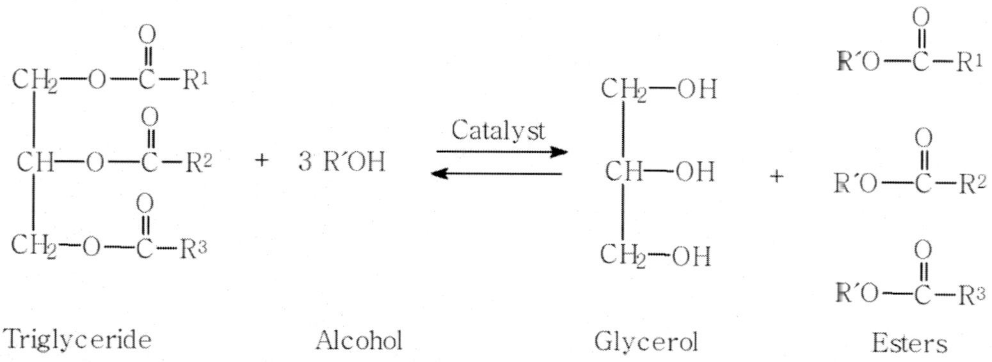

Figure 1. Transesterification of triglycerides with alcohol where R^1, R^2, R^3 are long-chain hydrocarbons, sometimes called fatty acid chains.

Transesterification is generally carried out by heating vegetable oils with excess alcohol under different reaction conditions in the presence of a catalyst. The reaction is reversible and therefore excess alcohol is used to shift the equilibrium to the products side. Alcohols used in the transesterification process are methanol (most frequently used), ethanol, propanol, butanol, and amyl alcohol. An acid, a base or an enzyme often catalyzes the reactions to improve the reaction rate and yield. Alkali-catalyzed transesterification is much faster than acid-catalyzed transesterification and is less corrosive to industrial equipment, and therefore is most often used commercially [14].

1.4. Homogeneous Catalysis

Transesterification of JC oil can be done in the same phase with the catalyst. Basic catalysts such as NaOH, KOH, and LiOH are used with an alcohol in the same phase with JC oil. In general, NaOH is preferred because of its low cost and its wide use in large-scale processing [15]. The alkaline catalyst concentration used in the reaction varies between 0.5-1.3% by weight with a yield of 94-99% conversion of JC oil into esters. The main advantage in using an alkaline catalysis process is the high conversion levels of triglycerides to their corresponding methyl esters in short reaction times, but the process is energy intensive and recovery of glycerol is difficult. In addition, the alkaline catalyst has to be removed from the product and the alkaline wastewater generated requires treatment. Levels of free fatty acids and water can also greatly interfere with the reaction because contamination results in soap formation that makes the separation process difficult [14, 15].

Homogeneously catalyzed transesterifications with acid catalysts have also been reported. Acid catalysts used in the process include H_2SO_4, HSO_3Cl, HCl, and H_3PO_4, [15, 20-22]. Based on the reaction rates, most of the reactions yielded at low to moderate efficiency at rates approximately 4000x slower than alkaline catalysis with minimum completion times of $\geq$12 h [15].

1.5. Heterogeneous Catalysis

As mentioned earlier, one of the disadvantages of homogeneous catalysis during the process of transesterification of vegetable oil is the removal of the catalyst. Several authors propose to solve this problem with the application of catalysts in heterogeneous phase using different compounds like CaO, BaO, MgO, ZnO, SrO, TiO_2, ZrO_2, La_2O_3, V_2O_5, $CaCO_3$, K_2CO_3, Na_2CO_3, $Mg(OH)_2$, and $Ca(OH)_2$ [15, 23-25].

The use of binary base catalysts has been also reported. These binary catalyst systems include the use of binary metal oxides, such as CaO-ZnO and CaO-La_2O_3 [25]; and a mixture of ZnO/zeolite [26] or PbO/zeolite [27]. The combination with zeolites was used with JC oil and both catalysts exhibit reasonably good activity for the reaction of interest and are reusable under the reaction conditions. In addition, the use of a combination with zeolites is desirable because it avoids leaching of metal ions during the reaction.

The synthesis of fatty acid methyl esters (FAME) using JC oil as the raw material in the presence of a binary catalyst system, such as CaO-NiO or CaO-Nd_2O_3, has been reported [28]. In this report, a FAME yield > 80% was obtained using a methanol/oil molar ratio of 15:1 in presence of the oxide catalysts (5%) at 65°C [28]. In addition, a combination of catalysts based on lanthanide oxides, such as Nd_2O_3-La_2O_3, yielded 93% of FAME using a methanol/oil molar ratio of 45:1 and a catalyst amount of 3 wt.% at 210°C over 4 h [29].

The synthesis of biodiesel from JC oil using heterogeneous nano-sized $CaTiO_3$ as a catalyst was studied in detail and its catalytic activity was compared with micro-sized $CaTiO_3$. Nano-sized $CaTiO_3$ afforded much higher yields of biodiesel as compared to micro-sized $CaTiO_3$. Furthermore, the nano-sized $CaTiO_3$ could be recycled just as the normal $CaTiO_3$. Easy preparation methods for nano $CaTiO_3$ can lead to new industrial procedures for large-scale preparation of biodiesel [30].

CaO with Li doping showed increased conversion to biodiesel than bare CaO as a catalyst. La_2O_3-ZnO, La_2O_3-Al_2O_3 and $La_{0.1}Ca_{0.9}MnO_3$ catalysts were also tested and among them La_2O_3-ZnO showed higher activity. Mixture of solid base catalysts (CaO and Li-CaO) and solid acid catalyst ($Fe_2(SO_4)_3$) were found to give complete conversion to biodiesel in a single-step simultaneous esterification and transesterification process [31].

1.6. Lipase Catalysis

Obtaining biodiesel by biotechnological methods has attracted much attention in recent years. The most widely used process is enzymatic transesterification to produce high-purity esters allowing easy separation of the glycerol byproduct. Lipases are able to hydrolyze lipids and then catalyze transesterification reactions. Lipase can be obtained from microorganisms such as *Mucor miehei, Rhizopus oryzae, Candida antarctica, Pseudomonas cepacia*, and *P. fluorescens* [14].

The advantages of using lipases are: (1) the regeneration and reuse of the immobilized enzyme, because it can be left in the reactor as long as a continuous flow of the reagents is maintained; (2) the immobilization of a high amount of lipases that allows longer activation of the enzymes; (3) higher thermal stability of the enzyme due to its native state; (4) the immobilization of lipases protects them from the solvent, preventing agglutination of the enzymes; and (5) the separation of the final product is easier when using this type of catalyst.

However, some disadvantages include: (1) lipases can lose some initial activity due to volume of the oil molecule, (2) the number of enzymes immobilized to the matrix is not uniform, and (3) immobilized enzymes are more expensive that the native enzyme [32].

Biodiesel production from JC seed oil by immobilized lipase-catalyzed transesterification using different bacterial and fungal sources is detailed in Table 3. A 100% yield was obtained with lipases from *P. cepacia*, whereas a combination of fungal lipases from *R. oryzae* and *C. rugosa* had a biodiesel production with conversion yield of about 80% [33].

1.7. Fuel Properties of *Jatropha* Biodiesel

The yield of biodiesel in the process of transesterification is affected by several process parameters, which include the presence of moisture and FFA, the reaction time, the reaction temperature, the catalyst, and the molar ratio of alcohol and oil. The finished biodiesel must be analyzed using established methodologies to ensure that it meets international standards. A few specifications have been formulated, but the (ASTM) D6751 and European Standard (EN) 14214 standards are most commonly used.

Table 3. Enzymatic production of *Jatropha* biodiesel using lipases

Lipase source	Alcohol	Solvent	Temperature (°C)	Yield (%)	Reference
Candida antarctica	Methanol	t-Butanol	35	88	[34]
Rhizomucor miehei	Methanol	None	45	81	[35]
Idiomarina sp. (Free)	Methanol	None	60	84	[36]
Idiomarina sp. (immobilized)	Methanol	None	60	91	[36]
Rhizopus oryzae and Candida rugosa	Methanol	None	45	90	[33]
Pseudomonas cepacia	Ethanol	None	50	98	[37]
Candida antarctica	2-Propanol	Hexane	50	93	[38]
Rhizopus oryzae	Methanol	None	30	80	[39]
Chromobacterium viscosum	Ethanol	None	40	92	[18]
Burkholderia cepacia	Ethanol	None	35	100	[40]
Enterobacter aerogenes	Methanol	t-Butanol	55	94	[41]

To ensure safe operation in diesel engines, the most important aspects of biodiesel product are the completion of the reaction, and the removal of residual catalyst, alcohol, and free glycerol. If the transesterification reaction is not complete then triglycerides, diglycerides, or monoglycerides may be left in the final product. Therefore, the ASTM standard requires total glycerol to be < 0.24% of the final biodiesel product. On the other hand, since residual alcohol, even as little as 1%, can lower the flashpoint of the final biodiesel product from 170°C to < 40°C, the EN 14214 standard limits the amount of alcohol to a very low level (< 0.2%). Finally, although a specific value for the residual catalyst is not included in the ASTM standards, it is limited by specifications on levels of sulfated ash, which may lead to engine deposits and high abrasive wear levels [16].

The fuel properties of *Jatropha* biodiesel are summarized in Table 4. *Jatropha* biodiesel has comparable properties with those of fossil biodiesel and conforms to international standards for biodiesel ASTM D6751 [42].

Table 4. Specifications and test methods ASTM D6751 and biodiesel properties

Property	Limits	*Jatropha* biodiesel	Test method	References
Acid value (mgKOH/g)	0.80 max	0.32-0.50	D664	[13-16, 45]
Cetane number	47 min	49-63	D613	[5, 13, 16, 43, 44]
Cloud point (°C)	report	1.3	D2500	[15]
Copper strip corrosion	No. 3 max	1a	D130	[15, 45]
Flash point (°C)	130.0 min	133-170	D93	[13-15, 44, 45]
Kinematic viscosity at 40°C (mm^2/s)	1.9–6.0	2.35-4.80	D445	[13, 15, 16, 43, 45]
Pour point (°C)	report	-6 to +6	D97	[13, 43, 45]
Specific gravity at 15°C (kg/m^3)	860-900	877-882	D287	[13, 15, 43-45]
Sulphated ash content % (m/m)	0.020 max	0.010-0.012	D874	[14, 45]
Free glycerol % (m/m)	0.020 max	0.018	D6584	[15]
Total glycerol % (m/m)	0.240 max	0.200	D6584	[15]

1.8. Advantages and Disadvantages of *Jatropha* Biodiesel

1.8.1. Advantages

Renewable Fuel. Reserves of non-renewable fossil fuels obtained from petroleum are rapidly declining, making the production of renewable fuels more attractive. *Jatropha* seeds can play an important role due to their high oil content suitable for biofuel production. *Jatropha* has wide distribution in the tropics, and it is found mostly at low altitudes ≤500 m above sea level, in dry or wet areas, plains or hills, with rainfall ranging between 300-1500 mm/annually, and optimum temperatures varying between 20-34°C. *Jatropha* is a drought resistant plant and it adapts to a variety of soils, including those with low nutrient content [46]. In addition, the plant can be used as a living fence around agricultural fields or on marginal soils to control erosion. Moreover, when the press cake is returned to the fields after oil extraction, there is sustainable recycling of nutrients and the soil remains productive [47].

Economic Development. *Jatropha* can be integrated into traditional farming systems in developing countries. The production and processing of seeds for biofuel provides extra job opportunities and can give local communities energy independence, and any excess of biofuel that is produced can be sold. The oil can also be used for soap production providing a profitable rural activity [47].

Cost Benefit. *Jatropha* biodiesel is a clean fuel, which means it produces fewer emissions on burning compared to fossil fuels. The biodiesel is also adaptable to current engine designs without any change, and it performs very well in most conditions. The "clean" burning of *Jatropha* biodiesel keeps the engine running for longer, lowers maintenance costs, and brings down overall pollution check costs when compared to "dirty" burning fossil fuels [48]. *Jatropha* biodiesel also has good brake thermal efficiency, which indicates the percentage of fuel energy converted to useful power output [49].

Air Pollution. Significant reduction of particulate matter and smoke emissions was measured when biodiesel is mixed with fossil fuels (fuel blends). This reduction is primarily a consequence of a complete combustion of the biodiesel, owing to its high oxygen content and its high cetane number [49].

Reduce Greenhouse Gases. Engines fueled with biodiesel reduce the emission of toxic gases such as CO, SO_2 and hydrocarbons [50].

Economic Security. Not all countries are petroleum producers and/or have large reserves. The economies of petroleum importer countries are greatly affected by the costs of fuel imports. Therefore an increase in the production of biodiesel, obtained from local *Jatropha* plantations, reduces reliance on imports. More jobs are created with a growing biofuels industry, which decreases dependency on other countries and strengthens the economy.

Specifications. *Jatropha* biodiesel has comparable fuel specifications with those from fossil diesel, and conforms to international standards for biodiesel (ASTM D6751, Table 4) [42].

Food Market Competition. Biodiesel obtained from *Jatropha* does not compete with the food market because *Jatropha* oil is not edible and lands dedicated to edible crops is maintained.

Net Energy Ratio (NER). Biodiesel contribution to non-renewable source depletion (mainly fossil fuels) depends on its net energy ratio (NER). NER is equal to biofuel energy content/ energy consumed for its production and distribution. In most of the biofuel production cases, NER is greater than 1. However, in some cases, NER is lower than 1, which means that these production systems are unacceptable from an energy point of view. NER depends on the limits of the system and other parameters, such as cultivation techniques, production methods, etc. [51]. For example, the energy benefits of biodiesel produced from *Jatropha* cultivation on two different types of soil (nutrient poor and normal soils) were evaluated with and without irrigation. The study concluded that energy values obtained regarding net energy balance and NER were high for both systems [52].

1.8.2. Disadvantages

NO_x Emission. It has been observed that the use of *Jatropha* biodiesel in a diesel engine increases NO_x emissions compared to fossil diesel. Biodiesel contains more oxygen than pure diesel and the complete combustion of biodiesels results in higher temperatures, avoiding oxygen dissociation, which eventually increases NO_x emission [49].

Production Cost. Due to the costs of catalysts, alcohol, and infrastructure; the price of biodiesel can vary from one region to another. Overall, biodiesel production is currently more expensive than fossil diesel [53].

Feedstock Cost. The growth and yield of *Jatropha* crop can be affected by many factors, such as climatic conditions, water availability, soil conditions, plant age, and pests and diseases [2, 53].

Use of Fertilizers. *Jatropha* crop could need fertilizer to increase or maintain production. The downside of using fertilizers is that they can have harmful effects on surrounding environment and may cause water pollution. Fertilizers contain nitrogen and phosphorus and can be washed away from soil to nearby natural water sources [2].

Water Use. Although JC is a drought-resistant plant still significant quantities of water are required to irrigate *Jatropha* crops for efficient yield. This may impose strain on local and regional water resources, if not managed wisely [2].

2. BIOGAS

Biogas is a gas generated artificially or naturally from fermentation of organic matter using biodigesters and bioreactors [54]. Biogas is constituted principally by methane (CH_4) carbon dioxide (CO_2), and may contain traces of hydrogen (H_2), hydrogen sulfide (H_2S), carbon monoxide (CO), siloxanes, halogenated hydrocarbons, and nitrogen (N_2). These tracers should be eliminated by treatment based on the use of the biogas [55].

Biogas is produce naturally in dumps, wetlands, and in the intestines of animals as a result of the microbiological activities of methanogenic bacteria.

Biogas as a combustible energy source is produced from biomass [56], and as mentioned earlier, it can be produced under controlled parameters using a biodigester (Figure 2).

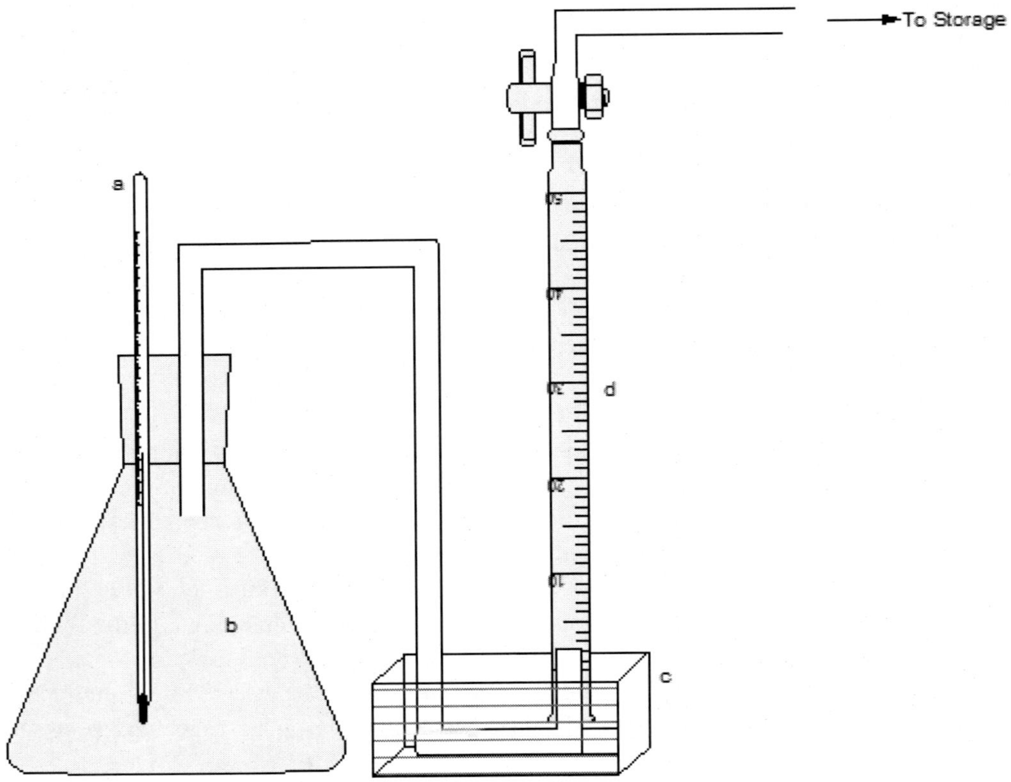

Figure 2. Scheme of a biodigester. (a) thermometer, (b) reactor containing the biomass to be anaerobically fermented, (c) container with water, and (d) inverted burette filled with water used to measure the gas produced.

The calorific value of a 1 m^3 of biogas is 6,000 kcal assuming a content of 70% methane and 30% CO_2. This value is comparable to the energy from 0.8 L of gasoline, 0.6 m^3 of natural gas, 1,5 kg of wood, 0.3 kg of charcoal, 0.71 L of fuel-oil, 1.2 L of alcohol, or 6.8 kWh of electricity [55].

In this section, a review of biogas production from de-oiled *Jatropha* residues is described.

2.1. Biogas from *Jatropha* Residues

As mentioned earlier, *J. curcas* is a bush that can grow up to 6 m and is adapted to different climates and soils without specific care [57]. In México, this plant, known as "piñón", produces toxic substances such as curcin (a toxic lectin) and phorbol esters that restrict its use as a source of food. In contrast, this plant is ideal for the production of biocombustibles because of the high oil content in its seeds. Moreover, this plant possesses multiple uses, such as in fertilizers, insecticides, and medicinal applications etc. as detailed in Scheme 1.

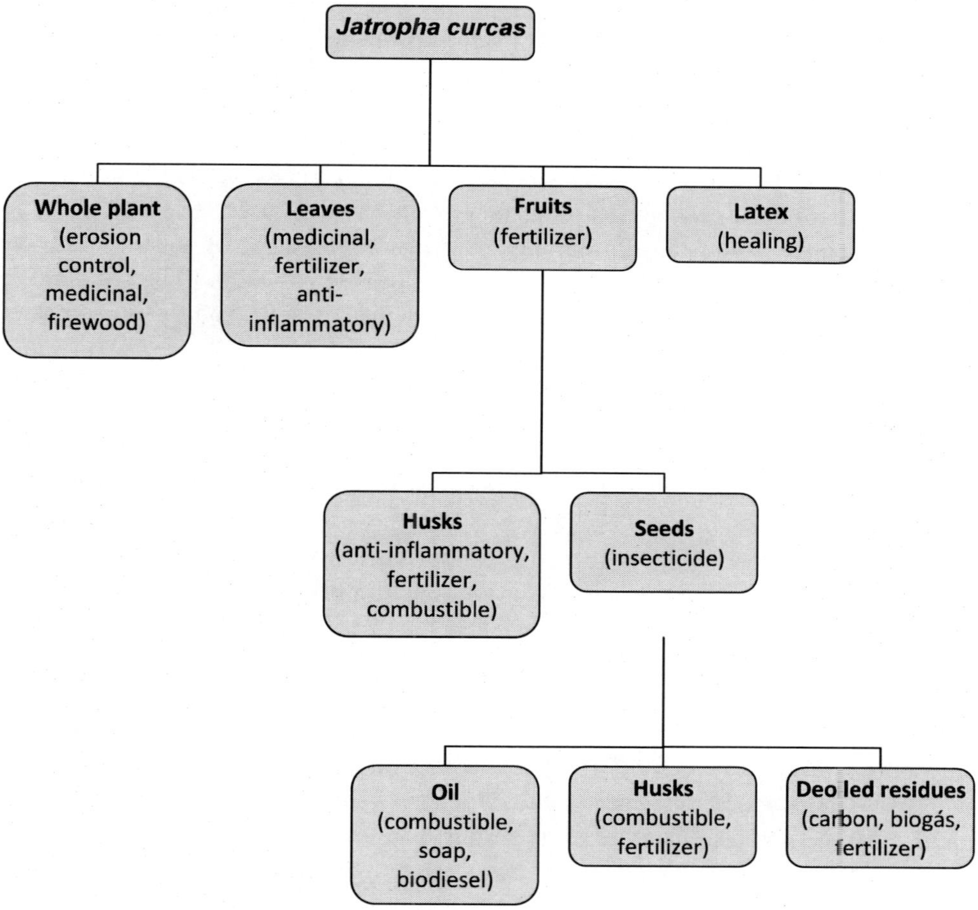

Scheme 1. Diversifications of *Jatropha curcas*. Adapted from [58-59].

2.2. Fermentable Material from *J. curcas* and Biogas Production

Anaerobic digestion consists of three main steps: (1) hydrolysis, (2) acidogenesis, and (3) methanogenesis. Different microbial consortia are involved in each of these steps and together transform the organic matter into mainly CH_4 and CO_2.

Total and volatile solids compose the organic matter used to feed the biodigester. Volatile solids are the materials that ultimately will be transformed into biogas, whereas total solids are mainly non-fermentable organic products that, together with recalcitrant minerals (e.g. K, Ca, N, and P), form the sludge or liquor, a mixture that can be used as a fertilizer. Then, although is desirable to have more volatile solids in the substrate for higher production of biogas, this claim should be taken carefully, because high content of solids favor the production of CO_2 instead of CH_4 [58].

Microbial consortia involved in the fermentation of organic matter include the following bacterial groups: hydrolytic-acidogenic, acetogenic, homoacetogenic, methanogenic, hydrogenofile methanogenic and acetoclastic methanogenic.

In order to efficiently produce biogas, a perfect balance between the different bacterial communities is desired. In addition, the velocity of the process will depend on the composition of the matter to be fermented and the composition and concentration of each microbial group.

In the case of *J. curcas*, de-oiled seed residues have been used as raw material for biogas production, representing approximately a 60% of the initial weight of the seeds. The biogas yield from the residues depends on what part of the seed is used. For example, higher production of biogas was obtained from the endosperm (798 mL g^{-1} of volatile solid) of the seed as compared to the episperm (216 mL g^{-1} of volatile solid) [60]. This difference in production is attributed to higher levels of fermentable matter in the endosperm compared to the episperm, which is mainly constituted by lignin [60].

Other parameters for biogas production to take into consideration are based on the physicochemical properties of the same de-oiled residue (seed cake). These parameters include moisture content and the levels of C, N, K, and P [61]. The control of these parameters is important for efficient management of the seed cake to avoid environmental problems.

Besides the characteristics of the raw material used for biogas production from JC, it is important to take into consideration other factors, such as pH, substrate-water dilution ratio, and the aforementioned bacterial consortia. In terms of pH, studies have reported that the highest biogas yield is obtained when the final pH is ≥ 8.0 with a low concentration of solids (seed cake to water ratio) [58]. The substrate:water ratio is important and varies between 1:6.67 to 1:10, but can reach up to 1:20 [58, 61-64]. Interestingly, an increase in the biogas production was observed when 5% (v/v) of a commercial mixture of 80 different microbial strains (EM-4) was added to JC husks. In this report, a substrate:water ratio of 1:8 reduced from 1:12 was used with similar yields, suggesting that this mixture can be used to reduce the level of water used for a biphasic fermentation system (hydrolysis and methanogenesis) [64]. Another study reported that a 25% increase in biogas production was recorded when JC cakes were mixed with cattle dung in a ratio of 25:75 [62]. Similar results were obtained when JC husks were mixed with cattle dung in a ratio of 33:67 [63].

In conclusion, the use of anaerobic digestion is beneficial because it reduces greenhouse emissions while producing energy from organic residues. Then, the benefits associated to anaerobic digestion are [55]:

- Significant decrease of odors associated with organic residues.
- Mineralization.

- Production of renewable energy.
- Emission decrease of gases related to the greenhouse effect.

3. BIOETHANOL

De-oiled seed cake from JC is rich in cellulosic material that can be used for bio ethanol production after fermentation.

Few studies have been published regarding the production of bioethanol from JC residues. All of these studies use a two-step process that includes acidic pre-treatment with diluted sulfuric acid and then chemical or enzymatic hydrolysis of the products before fermentation with *Saccharomyces cerevisiae*. The acidic pretreatment with diluted sulfuric acid (0.05%) and posterior hydrolysis enriches the sample with fermentable sugars with a concomitant increase in ethanol production after fermentation. A recovery of 80% ethanol was achieved with this process [65]. A modification of this process uses a slightly elevated concentration of sulfuric acid (1.5%) together with a thermal increase (136°C) in the pre-treatment, followed by enzymatic hydrolysis [66]. By using this process, authors were able to convert more than 80% of cellulose in the sample.

When growing JC plants, annual pruning is required. This pruning can give an average of 5 kg/year/plant of lignocellulosic material, which can be incorporated and economically converted into bioethanol [67].

4. LIFE-CYCLE ASSESSMENT (LCA) AND THE ENVIRONMENTAL IMPACTS OF INTRODUCING LARGE-SCALE PLANTATIONS

The need of renewable technologies to produce biofuels as an alternative to replace fossil fuels has led to the investigation of oleaginous plants. Studies have focused not only on the investigations of the oil content in plants, but also on the de-oiled residues. However, the use of extensive land to grow these plants indirectly increases the consumption of fossil fuels, fertilizers and pesticides with negative impacts on the ecosystem. In addition, the use of fertile soils for the cultivation of plants for energy production purposes decreases food production capacity. Taking together, the life-cycle assessment (LCA), or the assessment of the impact of a technology to the environment, does not currently favor using this technology, with standard biofuel crops (e.g. JC), as a complete replacement of fossil fuels. This is the reason that species like JC are under extensive study because they have low growth requirements which could potentially translate into less negative environmental impacts. The low requirements to culture this plant include: (1) adaptability to degraded and eroded soils with low water content; and (2) inedible with low competence for lands dedicated to crops for food production. However, studies evaluating the impact of life cycle and environmental impact of large-scale plantations of JC are scarce.

Taking together, the use of JC for biofuels production is a feasible alternative to fossil fuels. This fact is based on a LCA study on the production of biodiesel from JC. Results of this study demonstrated that the production and use of biodiesel from JC reduced the consumption of non-renewable energies by 82%. Other parameters calculated included the net energy ratio (NER) and the global warming potential (GWP). Values of 1.85 and 55% were calculated for NER and GWP, respectively. Interestingly, if the production of biogas is added, the NER or energy efficiency is improved to 3.4, whereas the GWP does not change significantly [68].

JC can be grown as a monoculture or associated to other crops. From an economical point of view, monoculture is attractive because production is higher, but it is detrimental to the ecosystem. Based on the adaptability of this plant, this issue could be negligible when growing this plant in degraded, eroded or uncover soils. Over the long term, these soils can be remediated and polyculture agriculture is more feasible. In this way, an optimization of resources is achieved taking into consideration that renewed soils are incorporated back into the culturing area and the fertility of the soils is maintained rather than exhausted by a monoculture practice. Interestingly, polyculture agriculture of JC with corn, beans and pumpkin has been reported [69]. In this study, the social, environmental and economic impacts were analyzed and a benefit of polyculture agriculture has been reported. However, more studies regarding focusing on improving production conditions are required [69].

In conclusion, the social, environmental and economic dimensions of JC biofuel production should be analyzed together in order to obtain reliable data addressing sustainability because an individual analysis of each dimension does not warrant a positive impact in the other two.

Acknowledgment

We thank Jeffrey Helm for helpful discussions.

References

[1] Islam AKMA, Yaakob Z, Ghani JA & Anuar N. *Jatropha curcas* L.: A future energy crop with enormous potential. In Hakeem KR, Jawaid M, Rashid U, editors. Biomass and Bioenergy: Applications. *Switzerland: Springer International;* 2014; 31-62.

[2] Li, Z.; Lin, B.; Zhao, X.; Sagisaka, M. & Shibazaki, R. (2010). System approach for evaluating the potential yield and plantation of *Jatropha curcas* L. on a global scale. *Environ. Sci. Technol.*, 44, 2204–2209.

[3] Islam, A. K. M. A.; Yaakob, Z.; Anuar N.; Primandari, S. R. P. & Osman, M. (2012). Physiochemical properties of *Jatropha curcas* seed oil from different origins and candidate plus plants (CPPs). *J. Am. Oil Chem. Soc.*, 89, 293–300.

[4] Akbar, E.; Yaakob, Z.; Kamarudin, S. K.; Ismail, M. & Salimon J. (2009). Characteristic and composition of *Jatropha curcas* oil seed from Malaysia and its potential as biodiesel feedstock. *Europ. J. Scient. Res.*, 29 (3), 396-403.

[5] Liu, Y.; Lu, H.; Liu, C. & Liang B. (2009). Solubility measurement for the reaction systems in pre-esterification of high acid value *Jatropha curcas* L. oil. *J. Chem. Eng. Data,* 54, 1421–1425.

[6] Min, J.; Li, S.; Hao, J. & Liu, N. (2010). Supercritical CO_2 extraction of *Jatropha* oil and solubility correlation. *J. Chem. Eng. Data*, 55, 3755–3758.

[7] Fernandez, C. M.; Fiori, L.; Ramos, M. J.; Perez, A. & Rodriguez, J. F. (2015). Supercritical extraction and fractionation of *Jatropha curcas* L. oil for biodiesel production. *J. Supercrit Fluids*, 97, 100-106.

[8] Kaul, S.; Sharma, G.; Porwal, J. & Bisht, N., (2011). Effect of low frequency ultrasonic assisted extraction on the quality of seed oils of Indian origin. *Fuel Proces. Technol.*, 92(10), 1813-1820.

[9] Shah, S.; Sharma, A. & Gupta, M. N. (2004). Extraction of oil from *Jatropha curcas* L. seed kernels by combination of ultrasonication and aqueous enzymatic oil extraction. *Biores. Technol.*, 96(1), 121-123.

[10] Sayyar, S.; Abidin, Z. Z.; Yunus, R. & Muhammad, A. (2011). Solid liquid extraction of *Jatropha* seeds by microwave pretreatment and ultrasound assisted methods. *Asian Network Scientif Inform.*, 11(13), 2444-2447.

[11] Sayyar, S.; Abidin, Z. Z.; Yunus, R. & Muhammad, A. (2009). Extraction of oil from Jatropha seeds-optimization and kinetics. *Am. J. Applied Sci.*, 6 (7), 1390-1395.

[12] Patil, P. D.; Gude, V. G. & Deng S. (2009). Biodiesel production from *Jatropha curcas*, waste Cooking, and *Camelina sativa* oils. *Ind. Eng. Chem. Res.*, 48, 10850–10856.

[13] Deng, X.; Fang, Z. & Liu Y. (2010). Ultrasonic transesterification of *Jatropha curcas* L. oil to biodiesel by a two-step process. *Energy Conv. Manag.*, 51, 2802–2807.

[14] Parawira, W. (2010). Biodiesel production from *Jatropha curcas*: A review. *Scientif. Res. Essays*, 5(14), 1796-1808.

[15] Sánchez-Arreola, E.; Martin-Torres, G.; Lozada-Ramírez J. D.; Hernández L. R.; Bandala-González, E. & Bach H. (2015). Biodiesel production and de-oiled seed cake nutritional values of a Mexican edible *Jatropha curcas*. *Renew. Energy*, 76, 143-147.

[16] Leung, D.Y.C.; Wu, X. & Leung, M.K.H. (2010). A review on biodiesel production using catalyzed transesterification. *Applied Energy*, 87, 1083–1095.

[17] Nakpong, P. & Wootthikanokkhan, S. (2009). Biodiesel production from mixtures of vegetable oil and used cooking oil. *Asian J. Energy Environ.*, 10(04), 221-229.

[18] Shah, S.; Sharma, S.; & Gupta, M. N. (2004). Biodiesel preparation by lipase-catalyzed transesterification of *Jatropha* oil. *Energy & Fuels*, 18, 154-159.

[19] Leung, D. Y. C.; Wu, X. & Leung, M.K.H. (2010). A review on biodiesel production using catalyzed transesterification. *Applied Energy*, 87, 1083–1095.

[20] Lemoine, G. & Thompson, R. W. (2014). A preliminary study of acid catalyzed transesterification of a *Jatropha*-like bio-oil. *Biomass Bioenergy*, 69, 169-174.

[21] Choudhury, H. A.; Srivastava, P. & Moholkar, V. S. (2014). Single-step ultrasonic synthesis of biodiesel from crude *Jatropha curcas* oil. *AIChE Journal*, 60(5), 1572-1581.

[22] Choudhury, H. A.; Srivastava, P. & Moholkar, V. S. (2013). Acid catalyzed biodiesel synthesis from Jatropha oil: Mechanistic aspects of ultrasonic intensification. *Chem. Engineer. J.*, 231, 262-272.

[23] Di Serio, M.; Tesser, R.; Pengmei, L. & Santacesaria E. (2008) Heterogeneous catalysts for biodiesel production. *Energy & Fuels*, 22, 207–217.

[24] Patil, P. D. & Deng S. (2009). Transesterification of *Camelina sativa* oil using heterogeneous metal oxide catalysts. *Energy Fuels*, 23, 4619–4624.

[25] Lee, H. V. & Taufiq-Yap, Y. H. (2014). Optimization study of binary metal oxides catalyzed transesterification system for biodiesel production. *Proc. Safety Environ. Protection*, DOI:10.1016/ j.psep.2014.10.001.

[26] Singh, D.; Ganesh, A. & Mahajani, S. (2014). Heterogeneous catalysis for biodiesel synthesis and valorization of glycerol. *Clean Technol. Environ. Policy*. DOI:10.1007/s10098-014-0858-9.

[27] Singh, D.; Rohidas, B.; Anuradda, G. & Sanjay, M. (2014). Synthesis of biodiesel from vegetable oil using supported metal oxide catalysts. *Energy & Fuels*, 28(4), 2743-2753.

[28] Hwa, S. T.; Umer, R. & Taufiq-Yap, Y. H. (2014). Biodiesel production from crude *Jatropha curcas* oil using calcium based mixed oxide catalysts. *Fuel*, 136, 244-252.

[29] Rabiah, N. M. F.; Taufiq-Yap, Y. H. & Hussein, M. Z. (2013). Production of biodiesel from non-edible *Jatropha curcas* oil via transesterification using Nd_2O_3-La_2O_3 catalyst. *Adv. Mat. Res.*, 620, 335-339.

[30] Surbhi, S.; Hait, S. K.; Arora, A. K. & Tuli, D. K. (2011) Transesterification of jatropha oil using heterogeneous nano sized $CaTiO_3$ catalyst. *J. Surf. Sci. Technol.*, 27(3-4), 173-191.

[31] Endalew, A. K.; Kiros, Y. & Zanzi, R. (2011). Heterogeneous catalysis for biodiesel production from *Jatropha curcas* oil (JCO). *Energy*, 36(5), 2693-2700.

[32] Marchetti, J. M.; Miguel, V. U. & Errazu A.F. (2007). Possible methods for biodiesel production. *Renew. Sustain. Energy Reviews*, 11, 1300–1311.

[33] Hyun, L. J.; Hah-Young, Y.; Young, J. S.; Sung, O. H.; Chulhwan, P. & Seung, W. K. (2012). Process development using co-immobilized lipases for biodiesel production. *Int. Proc. Chem. Biolog. Environ. Engin.*, 28, 171-174.

[34] Yingxia, L.; Dong, D. Y.; Yun, W. & Li, G. X. (2013). Response surface methodology optimization of lipase-catalyzed transesterification of *Jatropha curcas* L. seed oil for biodiesel production. *Biotechnol. Biotechnol. Equipm.*, 27(6), 4284-4289.

[35] Veny, H.; Aroua, M. K. & Sulaiman, N. M. (2014). Kinetic study of lipase catalyzed transesterification of *Jatropha* oil in circulated batch packed bed reactor. *Chem. Engin. J.*, 237, 123-130.

[36] Xin, L.; Po, Q.; Si-Guo, W. & Hui-Ying, Y. (2014). Characterization of an organic solvent-tolerant lipase from *Idiomarina sp.* W33 and its application for biodiesel production using *Jatropha* oil. *Extremophiles*, 18(1), 171-178.

[37] Shah, S. & Gupta, M. N. (2007). Lipase catalyzed preparation of biodiesel from Jatropha oil in a solvent free system. *Process Biochem.*, 42: 409-414.

[38] Modi, M. K.; Reddy, J. R. C.; Rao, B. V. S. & Prasad, R.B.N. (2006). Lipase-mediated transformation of vegetable oils into biodiesel using propan-2-ol as acyl acceptor. *Biotechnol. Lett.*, 28: 637-640.

[39] Tamalampudi, S.; Talukder, M. R.; Hama, S.; Numatab, T. & Kondo, A. (2008). Enzymatic production of biodiesel from Jatropha oil: A comparative study of immobilized-whole cell and commercial lipases as a biocatalyst. *Biochem. Eng. J.*, 39: 185-189.

[40] Rahmath, A. & Pogaku, R. (2013). Immobilized *Burkholderia cepacia* lipase for biodiesel production from crude *Jatropha curcas* L. oil. *Biomass Bioenergy*, 56, 8-13.

[41] Annapurna, K.; Paramita, M.; Kumar, G. V. & Rintu, B. (2009). Enzymatic transesterification of Jatropha oil. *Biotechnol. Biofuels*, 2:1, 1-7.

[42] ASTM. ASTM D6751-biodiesel blend stock specification (B100).

[43] Patil, P. D.; Gude, G. V. & Deng, S. (2009). Biodiesel production from *Jatropha curcas*, waste cooking, and *Camelina sativa* oils. *Ind. Eng. Chem. Res.*, 48, 10850–10856.

[44] Kumar, A. & Sharma, S., (2008). An evaluation of multipurpose oil seed crop for industrial uses (*Jatropha curcas* L.): A review. *Ind. Crops Prod.*, 28 (1), 1-10.

[45] Bhatnagar, A. K.; Kaul, S.; Chhibber, V. K. & Gupta, A. K. (2006). HFRR Studies on methyl esters of nonedible vegetable oils. *Energy Fuels*, 20, 1341-1344.

[46] Brittaine, R & Lutaladio, N. *Jatropha* cultivation. In: Integrated Crop Management Vol. 8–2010: *Jatropha*: A Smallholder Bioenergy Crop. The Potential for Pro-Poor Development. Rome: Food and Agriculture Organization of the United Nations; 2010; 27-39.

[47] Van der Putten, E. General data on *Jatropha*. In: Jan de Jongh, editor. The *Jatropha* Handbook: From Cultivation to Application. Netherlands: FACT Foundation; 2010; 1-7.

[48] Ghaly, A. E.; Dave, D.; Brooks, M.S. & Budge S. (2010). Production of biodiesel by enzymatic transesterification: Review. *Am. J. Biochem. Biotechnol.*, 6 (2): 54-76.

[49] Gaurav, P.; Ambarish, D. & Kumar, M. B. (2014). An experimental and numerical investigation of the performance, combustion and emission characteristics of a diesel engine fueled with *Jatropha* biodiesel. *Energy Procedia*, 54, 455-467.

[50] Pandey, A. & Parwani, A. (2014). Experimental analysis on biodiesel prepared from non edible oil resource to use in small agricultural diesel engine. *Int. J. Emerg. Technol.*, 5(1), 135-140.

[51] Sánchez-Arreola, E; Lozada-Ramírez, JD; Hernández, LR & Bach, H. Environmental Impact of Biofuels. In: Luis G. Torres & Erick R. Bandala Editors. Energy and Environment Nowadays. New York: Nova Science Publishers; 2014; 57-80.

[52] Chatterjee, R.; Sharma, V. & Kumar, S. (2012). Life cycle assessment of energy performance of biodiesel produced from *Jatropha curcas*. *J. Renew. Sustain. Energy*, 4, 053110. DOI: 10.1063/1.4754156.

[53] Johnston, M. & Holloway, T. (2007). A global comparison of national biodiesel production potentials. *Environ. Sci. Technol.*, 41(23), 7967-7973.

[54] Persson, S. P. E.; Bartlett, H. D.; Brandig, A. E. & Reagan R. W. (1979). Agricultural anaerobic digesters design and operation. The Pennsylvania State University, College of Agriculture, Agricultural Experiment Station, University Park, Pennsylvania. Bulletin 827.

[55] Instituto para la Diversificación y Ahorro de la Energía (2007). Biomasa: Digestores anaerobios. IDAE. Madrid, España. ISBN-13: 978-84-96680-21-0.

[56] Fang, C. (2010). Biogas production from food-processing industrial wastes by anaerobic digestion. PhD Thesis, DTU Environment Department of Environmental Engineering, Technical University of Denmark.

[57] Paramathma, M.; Parthiban, K. T. & Neelakantan K. S. (2004). *Jatropha curcas*. Forestry series-2 Publications, FC & RI, TNAU, Mettupalayam.

[58] Sinbuathong, N.; Munakata-Marr, J.; Boonsong, S. & Chulalaksananukul, S. (2011). Effect of the solid content on biogas production from *Jatropha curcas* seed cake. Proceedings of the Global Conference on Global Warming. 1-7. Lisbon, Portugal.

[59] FACT Foundation. (2010). The *Jatropha* handbook from cultivation to application. The Netherlands. FACT Foundation. 172 pp.

[60] Ryousuke, E.; Takahisa, A.; Toshio, S. & Yoshiaki, K. (2013). Methane fermentation of seed residues of *Jatropha curcas* after screw-pressed to obtain oil. *Baiomasu Kagaku Kaigi Happyo Ronbunshu*, 8, 46-47.

[61] Yadav, N.; Kumar, R.; Rawat, L. & Gupta, S. (2014). Physico-chemical properties of before and after anaerobic digestion of Jatropha seed cake and mixed with pure cow dung. *J. Chem. Engin. Proc. Technol.,* 5, 186. doi:10.4172/2157-7048.1000186.

[62] Nafisa, A.; Kurchania, A. K. & Swati, B. (2010). Bio-methanisation of *Jatropha curcas* defatted waste. *J. Engin. Technol. Res.,* (2), 38-43.

[63] Dhanya, M. S.; Gupta, N.; Joshi H. & Lata, C. (2009). Biogas potentiality of agro-wastes Jatropha fruit coat. *Int. J. Environ. Sci. Engin.,* 1, 136-140.

[64] Adinurani, P. G.; Setyobudi, R. H.; Wahono, S. K., Sasmito, A.; Nelwan, L. O.; Nindita, A. & Liwang, T. (2014). Optimization of concentration and EM4 augmentation for improving bio-gas productivity from *Jatropha curcas* Linn capsule husk. *Int. J. Renew. Energy Develop.*, 3, 73-78.

[65] Mohit, S. M.; Chandrashekhar, B.; Tanushree, C. & Kanwal, S. (2011). Production of bio-ethanol from Jatropha oilseed cakes via dilute acid hydrolysis and fermentation by *Saccharomyces cerevisiae*. *Int. J. Biotechnol. Applic.,* 3, 41-47.

[66] García, A.; Cara, C.; Moya, M.; Rapado, J.; Puls, J., Castro, E. & Martín, C. (2014). Dilute sulphuric acid pretreatment and enzymatic hydrolysis of *Jatropha curcas* fruit shells for ethanol production. *Ind. Crops Prod.*, 53, 148-153.

[67] Kumar, R.; Satlewal, A.; Sharma, S.; Kagdiyal, V.; Gupta, R. P.; Tuli, D. K. & Malhotra, R. K. (2014). Investigating Jatropha prunings as a feedstock for producing fermentable sugars and chemical treatment for process optimization. *J. Renew. Sustain. Energy*, 6, 033118/1-033118/11.

[68] Achten, W. M. J.; Almeida, J.; Fobelets, V.; Bolle, E.; Mathijs, E.; Singh, V. P.; Tewari, D. N.; Verchot, L. V. & Muys, B. (2010). Life cycle assessment of Jatropha biodiesel as transportation fuel in rural India. *Applied Energy*, 87, 3652–3660.

[69] Rucoba, G. A.; Munguía, G. A. & Sarmiento, F. F. (2012). Between *Jatropha* and poverty: reflections about biofuels production in temporary lands in Yucatan. *Estudios Sociales,* 21, 115-141.

In: Jatropha Curcas
Editor: Gregory Medina

ISBN: 978-1-63483-089-8

Chapter 2

TECHNOMIC ANALYSIS OF BIODIESEL PRODUCTION FROM *JATROPHA CURCAS*: SHRUB CULTIVABILITY, CONSECUTIVE-COMPETITIVE REACTIONS, CENTRIFUGAL SEPARATION AND OPTIMIZATION

Kal Renganathan Sharma*
Natural Sciences, San Jacinto College North Campus, Physics,
Texas Southern University, Houston, TX, US

ABSTRACT

Advances in sequencing of genomes has lead to the promise of genetically modified fuel crop such as *Jatropha Curcas* that is used to obtain Jatropha Oil. The gene regulation via Ribosome inhibiting Protein in the shrub has been studied using transfection of clones of curcin genes found in Jatropha Curcus and in other plants. This can lead to better cultivability of the shrub and higher yield of Jatropha Oil. NGS, Next-Generation Sequencing machines available commercially are tabulated. The triglycerides in Jatropha Oil can be transesterified using excess methanol in the presence of catalyst or excess methanol into biodiesel and glycerol. The glycerol can be sold as by-product. The kinetics of consecutive-competetive reactions during methanolysis of triglycerides with intermediates such as diglyceride, monoglyceride are modeled. The model solutions assuming irreversible simple first order kinetics were obtained using the method of Laplace transforms. Although biodiesel forms in each consecutive step under some conditions of reaction rate ratios ω, κ the yield of glycerol are seen to be higher than that of biodiesel. This can be called "cross-over" and can be seen in the illustrations shown as prototypical examples. The operating cost of separating biodiesel from glycerol using centrifuge can be obtained from computer simulations of layer formation during shear flow. For optimal total cost there exists a yield of biodiesel where the process can be operated at.

* E-mail: jyoti_kalpika@yahoo.com, kal.sharma@gmail.com.

Keywords: *J. Curcas*, biodiesel, gentically modified crop, gene regulation, consecutive-competitive reactions, centrifugal separation, transesterification, catalysis, optimization, oil well depletion, energy sustainability

INTRODUCTION

At a consumption rate of 84.6 million barrels per day the 237 trillion liters of crude oil [1a] reserves of the world can be expected to be depleted in another 74.5 years. The genominomics of *homo sapiens* and other species is seeing a cost reduction steeper [1] than the Moore's law for microprocessors according to the goal of NHGRI, National Human Genome Research Institute the cost per genome for humans is expected to decrease to less than $1000 by 2014 [2]. The cost of complete sequencing the genome of *J. Curcas* in order to look for valuable mutations is down to $50 in 2014. According to agricultural biotechnology company, SGB, the cost of sequencing *J. Curcas* five years ago was about $150,000. SGB spent $250,000 to create a master J. Curcas genome. With advances in transfection and genetic modifications it is fully expected that genetically modified *J. Curcas* will be used as a cash crop in order to provide feedstock for the burgeonoing biodiesel industry. *Jatropha oil* is non-edible [3]. Being non-edible use of *Jatropha oil* as feedstock for biodiesel production is not a fuel for food swap. This will increase public confidence in fuel use. *Jatropha* comes from the Greek words '*jatros*' that means doctor and '*trophe*' that means nourishement.

The genetically modified crop portion of the biotechnology industry is about $173.2 billion dollars in 2011. This includes the seeds, maize, soybean grain and cotton markets. The market share of Monsanto in seeds is the largest in the industry. SBG is growing hybrid strains of the shrub that can be used in order to produce biodiesel and by-product glycerol in quantitites that are comparable to petroleum when priced at $99 a barrel. This stain is called *J2*. SBG plans to grow the crop in 250,000 acres of *J. Curcas* in Brazil, India and other countries world wide. Dehgan [4] had studied the morphology of *J. Curcas* 30 years ago. India consumes more diesel than gasoline every year, i.e., about 320 million barrels of diesel every year compared with 94 million barrels of gasoline every year. This is in part because of the Indian Railways caters people travelling year around. Mass production of biodiesel is round the corner. Fairless [5] discuss the million *J. Curcas* seedlings planted on railway wastelands. *J. Curcas* is a member of the Euphorbia family that first was discovered in Central America, It has been used as lamp oil and soap. Out of the 756 million acres of land discussed in the report [5] by Ministry of Rural Development of Government of India, 56.5% of the land is already under cultivation and the rest is non-arable wasteland. This wasteland is used for cultivation of *J. Curcas* Shrub. The first commercial application of the *J. Curcus* shrub was reported in Lisbon. It was used as lamp oil and for production of soap. It thrives in tropical and sub-tropical regions of the world such as Africa and Asia. Portuguese ships were used in order to import *Jatropha Oil* [6]. *J. Curcas* is a large shrub and can survive for 50 years. It can attain a height of 9.44 m as discussed in the review of biodiesel production from *J. Curcas* by Addulla et al. [7] and others [8-10]. It can grow with less water in semi-arid conditions. It can grow on soils with less nutrient contents. Leaves are even toxic and is non-edible. Benefits of *J. Curcas* plant parts, use of wastelands, treatment needed prior to animal feed use, medicinal applications are discussed elsewhere [7]. *Jatropha Oil* contains about

25% protein, 47% fat and 5.5% moisture. It contains poly unsaturated linolenic acid and unstaturated linoleic acid in larger proportions. Seeds in Africa [11] were found to have 80% of linoleic acid and the seeds in India were found to have 81.9% linoleic acid ($C_{18}H_{32}O_2$). *Jatropha Oil* is extracted from the seeds using mechanical expelling of enzymatic methods [8]. The FFA content of *Jatropha Oil is* 14% which is higher than the 1% FFA content limit for use of base catalysts for conversion into biodiesel. Soap will form.

BIODIESEL PRODUCTION FORECAST

Biodiesel is a EPA, Environmental Protection Agency, designated advanced biofuel. It is a mixture of FAME, fattyacid methyl ester. It is increasingly used as an alternate energy fuel source of choice. The world food production is high enough for a food and fuel portfolio in some advanced nations in the world. The feedstock for biodiesel production can be Sunflower oil, Jatropha oil, Coconut Oil, Coconut Biomass, Waste Vegetable Oil, Soyabean Oil and Fats from Animal Husbandry. Catalytic transesterification of trigylcerides into diglyceride and then into monoglyceride and then into glycerol and FAME takes place in a set of consecutive-competitive reactions. The catalyst can be alkali, acid or enzyme. It depends on the FFA, free fatty acid content in the feedstock.

Over the past decade the world production of biodiesel has gone up from 15,200 barrels per day in the year 2000 to 300,000 barrels per day in 2010. In terms of volume this is about 5 billlion gallons in 2010. In the past two years, biodiesel production has exceeded targets. Production plants are present in nearly every state. 1000s of jobs are created The sigmoidal growth of the biodiesel volume is seen from 2006. Biodiesel is designated by ASTM D 6751 [12]. New laws and mandates on biodiesel came about in Brazil, China, United States, Argentina. Germany and Brazil are the world's leading biodiesel producers. Federal excixe tax credits are provided for producers and distributors of agri-biodiesel at $1 for every gallon of biodiesel they blend with regular diesel.

Forecasts of global dynamics of biodiesel production [13] are available for 2015-2020 by feedstock used such as vegetable oil feedstocks, jatropha oil, algae biodiesel and cellulose. The expected growth rate of biodiesel production in the world is about 6% between 2009-2018 according to OECD, Organization for Economic Cooperation and Development. By 2017 biodiesel production is expected at 25 billion liters. European biodiesel board estimated that the production of biodiesel in EU is about 9.6 milllion tons in 2010. By the year 2022, biofuel production is projected to consume a significant amount of total world production of sugar cane (28%), vegetable oils (15%) and coarse grains (12%).

In India, the former President of India, ABJ Abdul Kalam during his address to the nation on National Science Day, Feb' 28th 2006 called for an increase in output of biodiesel from J. Curcas crop from current levels of 2 tons per hectare to 4-6 tons per hectare [14]. The oil content of most jattropha varities range from 25-35%. Research in selection, intra-specific, inter-specific hybridization and mutation breeding is needed to develop varieties with more than 45% oil content so that a recovery of 35% under mechanical expelling is achieved. India has 60 million hectares of wasteland of which 30 million hectares are available for energy crops such as jatorpha. Cars that can run on biodiesel need be developed and encouraged. The

Indian Railways runs passenger trains with diesel engine with 5% blend of biodiesel. 15 million J. Curcas saplings are planted in Railways' land.

President B. Obama as a senator endorsed the budding biodiesel industry at a new biodiesel plant in Cairo, IL in 2006 [15]. The Renewable Energy group announced that it would build a 60 million gallon per year refinery and had raised $100 million in financing. Bunge Ltd., a major food processor and other venture capital firms were the contributors. About 76 biodiesel plants were in production in 2006, up from 22 in 2004. A biodiesel plant on an average costs up to $20 million to build and yields 30 million gallons per year of fuel. Biodiesel serves an important need of meeting the energy security of United States and the developing countries in the world.

Oil reserves are expected to be depleted by the year 2071 at the current levels of production. The crude oil reserves are estimated at 4.16 trillion liters world wide [16]. Global consumption is 84.6 million barrels/day. Earth's entire oil reserves according to one estimate is 1.2 trillion barrels without oil sands and 3.74 trillion barrels with oil sands. At the present rate of consumption the oil reserves will be depleted in the next 38.8 – 122.2 years. Search is on for alternative oil finds. Per geological survey 3-4.5 billion barrels was found in Montana and North Dakota. If oil shale can be used as source of oil the reserves can last for 110 more years. Oil finds have been found in Russia, Columbia and Africa. According to the big rollover theory global oil production is already past its peak production [17]. M. K. Hubbert, Shell Oil Co., Houston, TX, studied the exhaustion of oil fields. Initial oil find, exploitation and exhaustion phases were identified. This followed the bell curve. He concluded that United States would peak in its oil production in 1970. The curve is called the Hubbert cruve. The peak is also called the rollover. Lot of world oil experts feel that we are past the peak production. Every year since 1970 we have found less oil and pumped less oil. Air pollution has been found as a result of continued and increased use of petroleum. Global warming has been concluded as a problem because of significant increase in concentration of CO_2 in the earth's atmosphere [18]. The principles of Sustainable Engineering were developed at the Sandestin Conference of 2003 [19]. This ought to set the direction of engineers who work on developing sustainable alternatives to current engineering practices. Energy is considered a primary component of sustainable engineering. Biodiesel is nontoxic. It has low emission profiles and is environmentally benign [20]. A century ago R. Diesel successfully used vegetable oil as fuel for his engine. Prior to WW II vegetable oils were blended with diesel fuels time and again.

Centrifugal separation is used to separate the glycerin and biodiesel layers by gravity differences. More degree of separation can be achieved by increased torque of the rotor [21]. A trade-off is seen between utility cost for rotor speed and purity level. Optimal operation of rotor can be derived at for maximum revenue.

At end of the second stage [22] with 99.2 – 99.6% conversion the mixture is passed through a vacuum distillation tower in order to separate the unreacted methanol, recover the sodium methylate catalyst and recycle the unreacted oil. B & P Process patented a process [23] to make biodiesel with less equipment, more yield and at a higher purity. They use a higher temperature than the boiling point of methanol and increased pressure of the reactor in order to keep the methanol from boiling. The centrifugal separator was made with perforated concentric cylinders. The separation process was effected in a counter-current manner. This makes the throughput higher and use less floor space. The glycerin passes through the rims and the biodiesel separates out through the axial region of the separator. The reaction is between the triglycerides present in the oil and methanol. There are different methods to make biodiesel. One is by transesterification either using catalyst, enzyme or catalyst free and others are by pyrolysis and physical blending and emulsion processes. The catalyst used can be alkali, acid or enzyme. When the FFA, free fatty acid content is greater than 1% the acid catalyst would be better [24]. Alkaline catalyst is used in commercial plants. Alkaline catalysts are preferred when the FFA, free fatty acid content in the feedstock is less than 0.5 wt%. Process is sensitive to water and FFA. Saponification of ester may occur in presence of water. Pyrolysis methods have been found to result in more biogasoline compared with biodiesel [25].

NEXT-GENERATION SEQUENCING

Some of the NGS machines availabe for purchase or approved are shown in Table 1.0 [2]. The size of the device ranges from a small hand held device to bread-loaf size to desktop to bench top to a set of 10 consoles that can fill a room. The cost per genome or per read length can be calculated. The capital cost of the equipment can be amortized and added to the reagents cost, labor, supplies and overhead costs. Different concepts are used during the measurement of the sequence distribution.

The electrical current changes due to the polarity of the base pairs as DNA is passed through nanpore is one concept used. Another concept uses the change in electrical current as ions are released when the DNA reacts. The capital equipment cost can run to the order of multi million dollars. It can also be in the $1000 range depending on the read expectations. Sequencing by synthesis methods is used usually. Although a surragote Xpandomer is formed by encoding the ssDNA and as the Xpandomer passes through the nanopore it gets read. Run time is on the decrease across the board. Sample input sizes are to the tune of a few nanograms. Shot gun sequencing is used. Accuracy is improved. he variety of eukaryotes and prokaryotes that can be sequences are several ranging from bacteria to mammals. Some methods use lasers and optical devices and others use non-optical methods. The sequence information can be used in different applications. Disease specific catridges are developed by one vendor.

It can be used for personalized medicine or developing drugs for spreacing pandemics such as the recent Ebola virus scare. Electronic circuitry is synergized with biochip in some cases. Nanotrenches allow for efficient diffusion for probe target hybridization using fluorescent tags.

GENETIC MODIFICATION AND CULTIVABILITY

DNA fragments are isolated, inserted into what are called vector molecules and introduced into bacteria or yeast cells whey are allowed to replicate. High throughput DNA sequence analysis has resulted in completion of several genomes of eukaryotes and prokaryotes. Comparisons of sequences from among different organisms result in the realization of certain coding sequences and their functional metabolic products. DNA Microarrays are used to study metabolomics.

Table 1.0. Next Generation Sequencers Commercially Available

Name of Company	Device Name	Size of the Device	Features of the Device	Cost
Illumina, San Diego, CA	HiSeq X Ten System	10 Consoles of Ultra-high Throughput Sequencers	Can sequence 18,000 human genomes per year; can generate 1.8 trillion base pair data; run time less than 3 days	$10 million;
	NextSeq 500	Benchtop	3 billion base pairs can be sequenced in a day	Per genome price less than $1000 $250,000
hotlink			*www.illumina.com*	
Thermo Fisher Scientific	Sanger Sequencer		Read lengths of several hundred base pairs.	$50-$250
	ClaSeek and MuSeek Kits		5 ng – 1 μg sample input for ClaSeek kit and 100 ng of gDNA, cDNA for MuSeek kits. 120 min amplification time for ClaSeek and 80 min total time for MuSeek Kits. Build DNA library	$1,120
	IonPGM™ Hi-Q™ Sequencing Kit		200-400 base pair sequencing. Measures ions released as DNA reacts	
hotlink			*www.thermofisher.com*	
Roche	GS Junior and GS FLX System	desktop (55 lb weight)	Read length of 1000 base pairs, high accuracy and high throughput; Shot gun reads. Size of genome spectrum from bacteria to mammals; Compare sequence reads for SNPs; pyrosequencing	Goal- Ultra low cost
	Stratos Genomics' SBX Method		Single Molecule Detection. Single drop biochemical reaction to encode the sequence of DNA into a surrogate called Xpandomer. Xpandomer passed through nanopore and base pairs are read.	
hotlink			*www.roche.com*	
Pacific Biosciences	RS II System	300 lb benchtop	Uses lasers to sequence DNA; Average Read length is ~ 8500 base pairs; Single Molecule real time sequencing; DNA reactions are tracked in real time across 150,000 nanotrenches. Flourescent tags. Published a human reference genome.	$700,000
hotlink			*www.pacificbiosciences.com*	
Oxford Nanopore	MinIon	Handheld, disposable	A linear ssDNA is passed through a nanopore and the changes in electrical current is used to deduce the base pairs.	$900
hotlink			*www.nanoporetech.com*	
Name of Company	Device Name	Size of the Device	Features of the Device	Cost
QuantuMDx, New Castle, United Kingdom	Q-POC™	Handheld, Chip Based	15 minute bedside diagnoses; Disease Specific Catridges; Nanowire Biosensors (under development); Infectious Diseases; Tumor Profiling	$5- $20
hotlink			*www.quantumdx.com*	

Name of Company	Device Name	Size of the Device	Features of the Device	Cost
GenapSys Redwood City, CA	GENIUS – Gene Electronic Nano-Integrated Ultra-Sensitive Technology	Bread-Loaf Sized	Measures electric signal from DNA as it is copied. Code deduced from electrical signal	Costs a few thousand dollars; $50 genome and point-of-care diagnostic use
hotlink			*http://genapsys.com*	

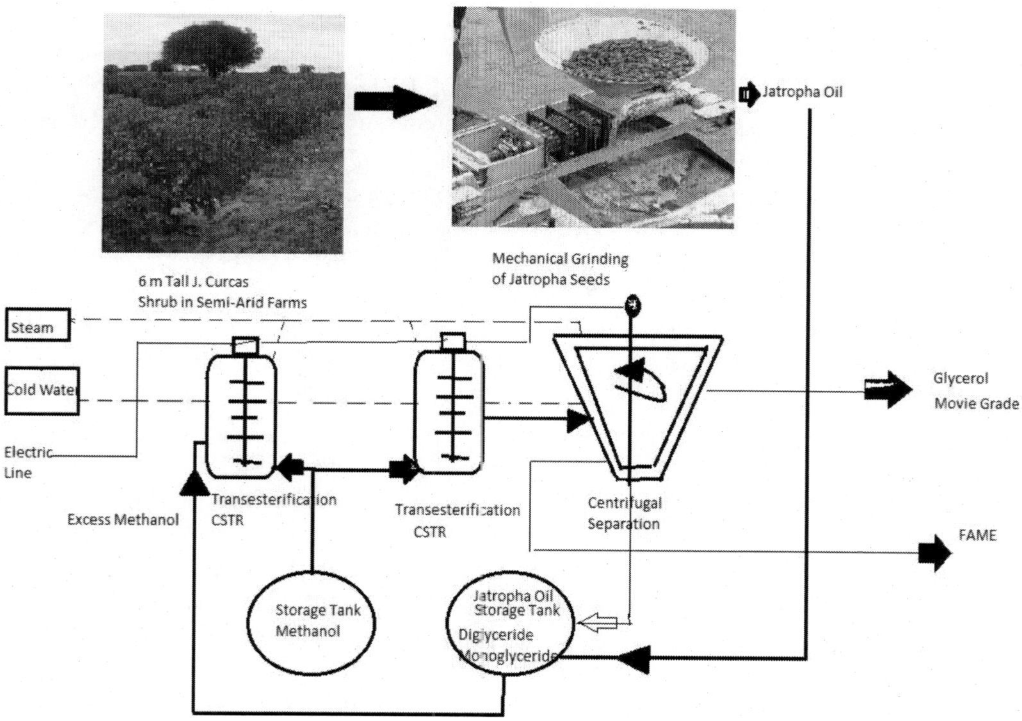

Figure 1a. Flow Diagram for Continuous Production of Biodiesel and Glycerol from J. Curcas Shrub Fuel Crop in Semi-Arid Farms.

Here global patterns and coordinated regulation of gene expression are investigated. Protein –protein interactions can be detected using two-hybrid analysis.

Transgenic, cisgenic and subgenic are different methods of gene transfer. Transgenic method of gene transfer is when genes from other species are inserted into the plant. Cisgenic transfer is when the genes are transferred to the plant from the same species or closely related species. Subgenic transfer is gene modification by using gene editing tools such as CRISPR and TALENS.

The genetically modified crop portion of the biotechnology industry is about $173.2 billion dollars in 2011. This includes the seeds, maize, soybean grain and cotton markets. The market share of Monsanto in seeds is the largest in the industry.

Qin et al. [26]. investigated the regulation expression of β-glucuronidase, GUS reporter gene in *Nicotiana Tabacum* by curcin promoter. This was prepared from *J. Curcus L* endosperm by cloning. A 0.6 kb fragment of a 5' flanking region adjacent to the curcin gene of *J. Curcas* endosperm was cloned. This fragment encodes a Type I, Ribosome-inactivating protein, RIP. The fragment of the curcin gene was used in order to drive the gene expression of GUS reporter gene found in *Nicotiana Tabacum*. The promoter was found to be active in the endosperm tissue of dicotyledonous tobacco embryo. The activity was initiated at the embryonic stage during seed development.

Often times in the field of plant genetic engineering the expression levels of target gene in certain tissues of the modified plant is specified. Promoter studies are used in design of bioreactors in order to manufacture proteins. Identification of molecular elements that can be used in order to control expression of external genes inside plants and promoter activity study are salient considerations. RIPS may be used to inactive ribosomes and serve as inhibitor of protein production. Curcin gene expression patterns can be used in scale-up of therapeutic drugs. Curcin genes have been cloned from seeds of *J. Curcas*. Qin et al. [26]. collected *J. Curcas* seeds and germinated them in plastic pots containing nutrient soils. They were grown in the conservatory greenhouse for two months. *Nicotiana Tabacum* plants were grown at 25^0C under a 16 hour-8 hours of light and dark photoperiod regimes. For purposes of plant transformation *Escherichia Coli* and *Agrobacterium tumefaciens* stains were also used.

The 5’ flanking sequence with Genbank accession # AF469003 was amplified using two primers. The forword primer P1F had the sequence 5’CCA**AAGCTT**-AATATTGGAATAGA AGACTTTG3’ and the reverse primer P2R had the sequence 5’-CCA**GGATCC**-CAAATATCATTATACGAATACG3’. A HindIII site (AAGCTT) was added to the forword primer at the 5’ end and a BamHI site (GGATCC) was added to the reverse primer as shown in bold face. The amplified sequences were cloned into pMD18-T vector (TaKaRa) on both strands. The Curcin promoter was cut from pMD18-T vector with HindIII and Bam HI fragments. The recombinant plasmid pBI121-CPI formation is shown in Figure 1.0. The CaMV 35s promoter was replaced with the cloned HindIII-BamHI fragments. The plant expression vectors was transferred into cells of *Agrobacterium Tumefaciens* by freeze-thaw treatment. The gDNA was extracted from tissues in the leaf. PCR analysis was carried out.

The gDNA from *Nicotiana Tabacum* was digested with HindIII and separated using agarose gel electrophoresis and transferred onto nylon membrane. GUS gene fragment was amplified using PCR from pPI121 using primers. DIG High Prime DNA Labelling and Detection Starter Kit II from Roche, Germany was used for hybridization and immunological detection. Fluorogenic reaction was carried out in 2mM 4-methylumbelliferyl-n-L-glucuronide, MUG extraction buffer.

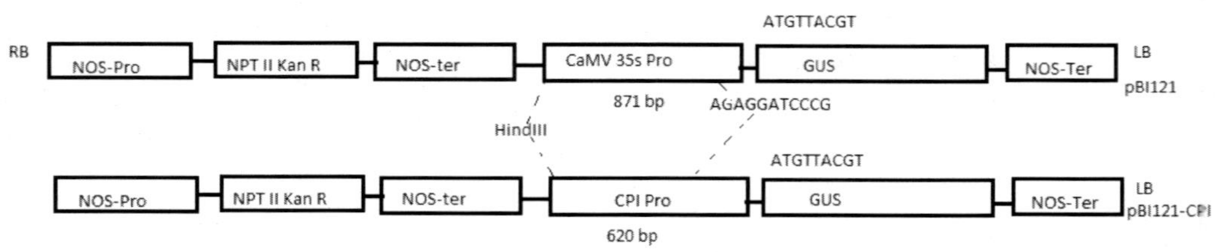

Figure 1.0. Recombinant Plasmid Formation – CPI Replaces CaMV 35s Promoter in Plasmid pBI121.

Fluorescence was measured using spectrofluorimeter with excitation beam at 365 nm and emission beam at 455 nm wavelengths. Different lengths of CPI were fused to the GUS reporter gene and transferred to tobacco plants. CPI promoter sequence was found to regulate expression of GUS gene in the endosperm seeds of *Nicotonia Tabacum*. Regulatory motifs contained in the endosperm that might have played a role in GUS gene expression a search of the PLACE database was conducted. -377 to -179 bp in the genome was found to be the endosperm responsive region. It contained four AAAG , one GT-1 binding site one E box and one W boc TTGAC motifs. AAAGsequence has been found by prior investigators for gene expression in maize that is endosperm specific. These bind to the Dof protein. Dof proteins are DNA binding proteins that are found in plants and can increase transcription. One Dof protein isolated from maize, PBF is bound to Prolamin box. WRKYs are involved in the regulation of the development of seed and trichomes and defense against pathogen infection.

Zhang et al. [27] applied next-generation Illumina Next-Generation Sequencing technology in order to study global gene expression patterns in leaves and roots and leaves of *J. Curcas*, 2 hours, 2 days and 7 days after the onset of salt stress. They found that 1504 genes were up regulated, 1115 genes were down-regulated in leaves and roots under salt stress condition. Gene ontology studies reveal that number of metabolic processes that occur in the plant are affected by salt stress. The genes were found to regulate ABA and ethylene signaling, osmotic regulation, the reactive oxygen species scavenging system and the cell structure in the leaves and roots. Salt stress was found to interfere with plant growth and production. The molecular mechanisms for the salt response in leaves and roots were attempted to be studied. The morphological adaptations of plant in response to abiotic stress may be linked to gene expression. Productivity of crops are hampered by salinity of the soil. Ionic and osmotic stresses emnate from saline environment. Transcriptome studies on plants exposed to salt stress have been undertaken for maize, cotton, flowering plants. *J. Curcas* is a perennial shrub and belongs to the species of *Euphorbiaceae*. It has high potential for biodiesel production. It is oil rich, drought-tolerant shrub. It is yet to be domesticated. Large scale plantation of *J. Curcus* would need conversion of this species into a genuine crop. Next-generation sequencing studies of *J. Curcus* may lead to attainment of these goals. It is a diploid with a haploid genome size estimated at 416 Mbp. Sato et al. [28] completed the sequencing of whole genome of *J. Curcas*. They used a combination of Sanger method and Next-Generation Sequencing method. They found 285,858,490 bp. The non-reduntant sequences comprised of 120,586 contigs and 29,831 singlets. They accounted for 95% of the gene-containing regions with the average Guanine + Cytosine contents of about 34.3%. 40,929 complete and partial structures of protein encoding genes have been inferred. Sequence comparison studies reveal that 4% or 1529 putative protein-encoding genes are specific to the *Euphorbiaceae* species. More microsynteny was found between *J. Curcus* and genome of castor bean and less microsynteny was found with soybean and *Arabidopsis Thaliana*. Pyrosequencing was used to characterize cDNAs extracted from tissues of the shrub. 21,225 unigene data was obtained. (http://www.kazusa.or.jp/jatropha). Genetic diversity was characterized by polymorphism analysis using microsatellite markers. Biofuel production is expected to increase from the understanding achieved from these studies. Breeding of the fuel crop was accelerated by genomic selection.

Silva-Junior [29] found a set of SNPs for *J. Curcus* using Illumina sequencing. Two Illumina GAIIx single end lanes were sequenced by using standard protocols. Raw reads were processed and aligned on mapped reference genome. Allele frequency was estimated by using Genotyper. A Bayesian genotype likelihood model was used in order to provide a posterior probability of occurrence of segregating variant allele at each locus. Illumina Golden Gate Genotyping Technology assays were designed using SNPs and in silico estimated minor allele frequency greater than 0.1. At least 60 bases were available on each SNP flank with no additional SNPs following. They report a unigene length of 39.7 Mbp and ~ 56% of the transcribed portion of the genome. They sampled 28,110 unigenes.

There was low seed yield found with *J. Curcas L* as result of unreliable flowering. Flowering Locus T, FT like genes are important flowering regulators in higher plants such as in *Arabidopsis thaliana* [2]. In order to better understand the genetic control of flowering in *J. Curcas* an FT homolog, JcFT was isolated and characterized by Li et al. [30]. They found sequence analysis and phylogenetic relationship with the FT genes of *Litchi Chinensis*, *Populus Nigra* and other perennial plants. JcFT may encode a florigen that may act as a key regulator in flowering pathway.

There are three critical areas in the continuous process for manufacture of biodiesel: (i) Feedstock Preparation; (ii) Consecutive-Competitive Reactions and; (iii) Separation of Biodiesel and Glycerol. The total cost of the process may be optimized with respect to capital cost and operating cost. The AW, annual worth [31] analysis of a biodiesel manufacturing plant for different feedstocks in Taiwan was discussed [32]. With by-product sales credit for glycerol the process may be profitable depending on the raw material cost and market price of gasoline. The cost of raw materials is a critical factor in the profitability of biodiesel production. Twelve reports were reviewed [33, 34] on economic feasibility of biodiesel production using different feed stocks and scales of operation. Significant factors that contribute to the bottom line of the biodiesel production were identified [35-38]. These include the cost of raw materials, plant size, credit received for glycerine as by-product sales. When waste cooking oil was used the material costs went down. Restaurant greases cost less than food-grade canola and soybean oils. The first factory that produced biodiesel at 300 tons per year from waste cooking oil was started in Chiayi county of Taiwan in October of 2004. The total cost of production can be written as follows;

$$\text{Total Cost} = \text{Capital Cost} + \text{Operating Costs} \tag{1}$$

The capital costs can increase with increase in reactor size needed to perform the reactions. The reactor size will be larger for larger reaction times or higher conversion targets. However the separation costs of separating biodiesel from glycerol, FAME mixture will be lesser at higher conversion from the reactors. When the conversion is lower the capital cost will decrease on account of the reactor size and the separation costs will increase on account of the load on the rotor of the centrifuge used. The variation of reactor size with conversion for a CSTR can be seen to be exponential for a given throughput. The variation of utility cost for a given throughput with conversion of *Jatropha Oil* in the reactors can be expected to be non-linear and can be obtained from computer simulations. Eq. (1) can be expressed in terms of conversion of *Jatropha Oil*. The resulting equation can be differentiated with respect to conversion and equated to zero and the extremamas can be obtained. The conversion corresponding to a minima can be obtained by confirming that the second derivative of the objective function is negative at the extremama. This gives impetus to study the kinetics of the consecutive-competitive reactions in the reactors and the velocity profiles in the centrifuge during separation of the product and by-product.

CONSECUTIVE-COMPETITIVE REACTIONS

It can be seen from the economic analysis the yield of biodiesel compared with other by-products such as glycerol can be a critical design criteria in making these plants more profitable. The reaction sequence for formation of FAME, faty acid methyl ester from triglycerides found in palm oil and other feedstock involve the formation of diglycerides, monoglycerides and glycerol in sequence with FAME produced in each intermediate step [39].

The reaction scheme can be represented as shown in Figure 2.0. The flow diagram for the biodiesel, glycerol factory from J. *Curcas* cash crop. is shown in Figure 1a. The reactions are catalyzed. The catalyst type depends on the FFA content in the feedstock. The triglycerides species is represented with symbol A, diglycerides with R, monoglycerides with S and glycerol with T. The product FAME formed in each step is given by P and the methanol used is given as B. Om Tapanes et al. [40] studied the reaction pathways and reaction sequences during base catalyzed transesterification of triglycerides of fattyacids including linoleic acid and determined the most probable pathway and the rate determining step of the reactions using molecular orbital calculations. The scheme in Figure 2.0 may be applicable for the biodiesel production from *Jatropha Oil.* The reactions in the reactor during biodiesel production may be modeled as scheme of multiple reactions of the consecutive-competitive/series-parallel type [41]. The methanol can be assumed to be in excess. Hence the reactions shown below can be assumed to obey the pseudo first order kinetics. The concentration of methanol can be lumped with the intrinsic second order reaction rate constant to give a pseudo first order lumped rate constant. The catalytic effect is also captured here. The reactions are modeled as follows;

The reactions in the reactor during biodiesel production may be modeled as scheme of multiple reactions of the consecutive-competitive/series-parallel type. The methanol can be assumed to be in excess. Hence the reactions shown below can be assumed to obey the pseudo first order kinetics. The concentration of methanol can be lumped with the intrinsic second order reaction rate constant to give a pseudo first order lumped rate constant. The catalytic effect is also captured here.

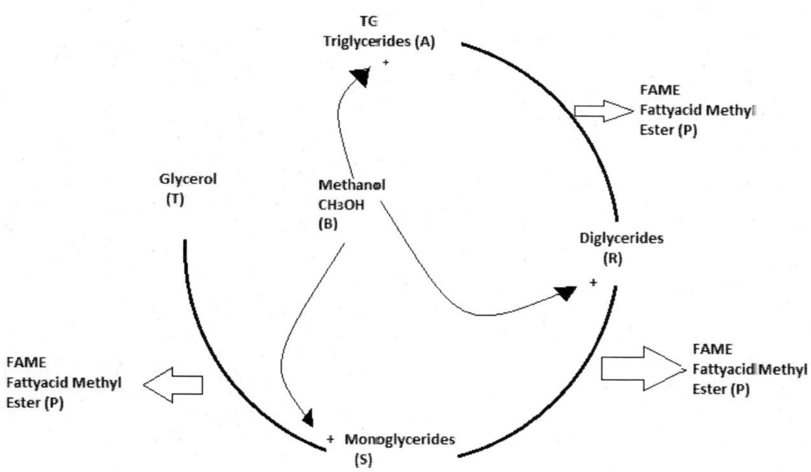

Figure 2.0. Transeterification Catalyzed Reactions From Triglycerides to Glycerol and FAME.

The reactions are modeled as follows;

$$A + B \xrightarrow{k_1} P + R$$
$$R + B \xrightarrow{k_2} P + S$$
$$S + B \xrightarrow{k_3} P + T \tag{2}$$

where A - triglyceride, B - Methanol (CH_3OH), R-1,2 and 1,3 diglyceride, S – mono glyceride, P – (FAME), T – Glycerol. It may be assumed that once the product P is formed it does not participate in the reaction any further. The FAME is harvested from the kettle. As glycerol (T) can be sold for profit this scheme is of more interest. This reaction set is applicable for successive attacks of a compound by a reactive material. In this case the reactive material is methanol and the compound is triglyceride. The kinetics of the reactions can be written as follows;

$$\frac{dC_A}{dt} = -k_1 C_A$$
$$\frac{dC_R}{dt} = k_1 C_A - k_2 C_R$$
$$\frac{dC_S}{dt} = k_2 C_R - k_3 C_S$$
$$\frac{dC_T}{dt} = k_3 C_S$$
$$C_P = C_{A0} - C_R - C_S - C_T - C_A \tag{3}$$

The scheme of reactions can be modeled as shown in Eq. (2) as a consecutive-competitive type [41]. The reaction rate expressions in Eq. (3) can be written in dimensionless form as follows after making the following substitutions;

$$X_A = \left(\frac{C_{A0} - C_A}{C_{A0}}\right)$$
$$X_R = \left(\frac{C_R}{C_{A0}}\right)$$
$$X_S = \left(\frac{C_S}{C_{A0}}\right)$$
$$X_T = \left(\frac{C_T}{C_{A0}}\right)$$
$$X_P = \left(\frac{C_P}{C_{A0}}\right)$$
$$\tau = k_1 t$$
$$\kappa = \left(\frac{k_2}{k_1}\right)$$
$$\omega = \left(\frac{k_3}{k_1}\right) \tag{4}$$

In dimensionless form the rate expressions given in Eq. (3) can be seen to become;

$$\frac{dX_A}{d\tau} = 1 - X_A \tag{5}$$

$$\frac{dX_R}{d\tau} = 1 - X_A - \kappa X_R \tag{6}$$

$$\frac{dX_S}{d\tau} = \kappa X_R - \omega X_S \tag{7}$$

$$\frac{dX_T}{d\tau} = \omega X_S \tag{8}$$

The rate expression for the product, FAME can be obtained by adding the contributions from the methanolysis of triglyceride, diglyceride and monoglyceride steps and can be seen to be;

$$\frac{dX_P}{d\tau} = 1 - X_A + \kappa X_R + \omega X_S \tag{9}$$

In order to evaluate the selectivity of the FAME product P over the by-product, glycerol, T solution to Eqs. (5-8) were obtained by the method of Laplace transforms [42]. The solutions are as follows;

$$X_A(s) = \frac{1}{s(s+1)}$$
$$X_A(\tau) = \left(1 - e^{-\tau}\right) \tag{10}$$

$$X_R(s) = \frac{1}{(s+\kappa)(s+1)}$$
$$X_R(\tau) = \frac{1}{(1-\kappa)}\left(e^{-\kappa\tau} - e^{-\tau}\right) \tag{11}$$

$$X_S(s) = \frac{\kappa}{(s+\kappa)(s+1)(s+\omega)}$$
$$X_S(\tau) = \frac{1}{(1-\kappa)(1-\omega)(\kappa)}\left((1-\kappa)e^{-\omega\tau} - (1-\omega)e^{-\kappa\tau} - (\kappa-\omega)e^{-\tau}\right) \tag{12}$$

$$X_T(s) = \frac{\omega\kappa}{(s)(s+\kappa)(s+1)(s+\omega)}$$

$$X_T(\tau) = \frac{\omega\kappa}{(1-\kappa)(1-\omega)(\kappa)}\left((1-\kappa)(1-e^{-\omega\tau}) - (1-\omega)(1-e^{-\kappa\tau}) + (\kappa-\omega)(1-e^{-\tau})\right) \quad (13)$$

The product yield can be found by difference as;

$$X_p = X_A - X_R - X_S - X_T \quad (14)$$

Model solutions given by Eqs. (10-14) were plotted in Microsoft Excel 2010 for Windows 7.0on a Hewlett Packard Compaq Elite 8300 desktop computer with Intel Core i7 processor with 3.9 GHz speed. The results for the product distribution is shown in Figures 3.0 – Figures 6.0. The simulations were conducted for values of reaction rate constant ratios κ < 1 and ω < 1 and further for ω < κ.

It can be seen from the Figures 3.0-6.0 that the conversion of species A, X_A increases in in a monotonic manner as predicted in Eq. (10).

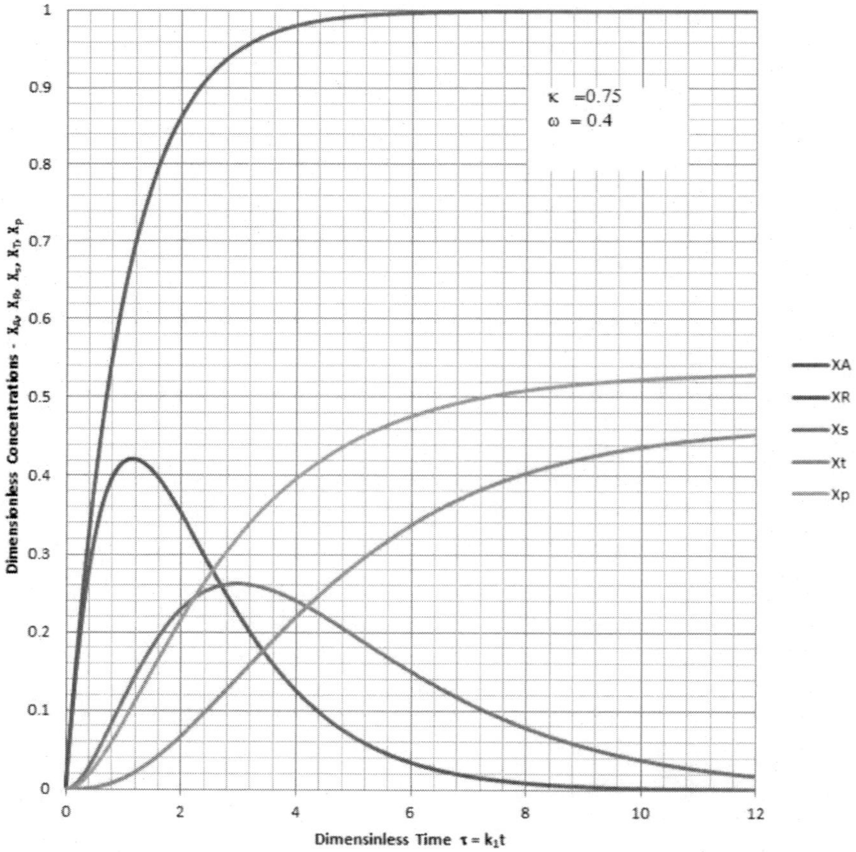

Figure 3.0. Trigylceride (A), Diglyceride (R), Monoglyceride (S), Glycerol (T) and FAME (P) Product Ditribution in Progressive Methanolysis[44] at κ = 0.75 and ω = 0.4.

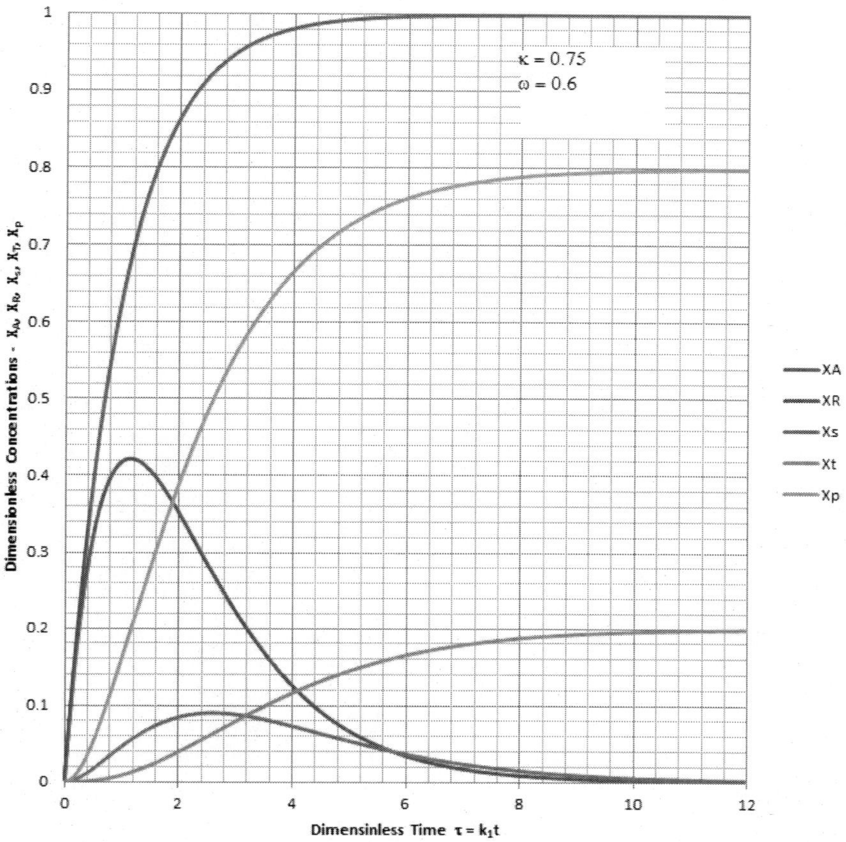

Figure 4.0. Trigylceride (A), Diglyceride (R), Monoglyceride (S), Glycerol (T) and FAME (P) Product Ditribution in Progressive Methanolysis at κ = 0.75 and ω = 0.6.

The monoglyceride and diglyceride yields go through a maxima. A change in curvature from convex to concave can be seen in the product yields of FAME and glycerol.

There is a rate increase later in time in the formation of glycerol. The selectivity of FAME can be poor compared with that of glycerol formation as can be seen in Figure 4.0. FAME yield can be high as shown in Figures 3.0, 4.0. There can also be a "cross-over" from higher selectivity of FAME to lower selectivity of FAME compared with glycerol as can be seen in Figure 6.0. In such cases, CSTR can be used with residence times less than the cross-over point in order to obtain higher yield of FAME. The convexo-concave curvature in the product yields is consistent with experimental studies such as that repored by Jaya and Ethirajulu [43].

Centrifugal Separation of FAME and Glycerol: Torque Requirements

The separation costs of FAME and glycerol may be a critical factor in the process design of biodiesel production.

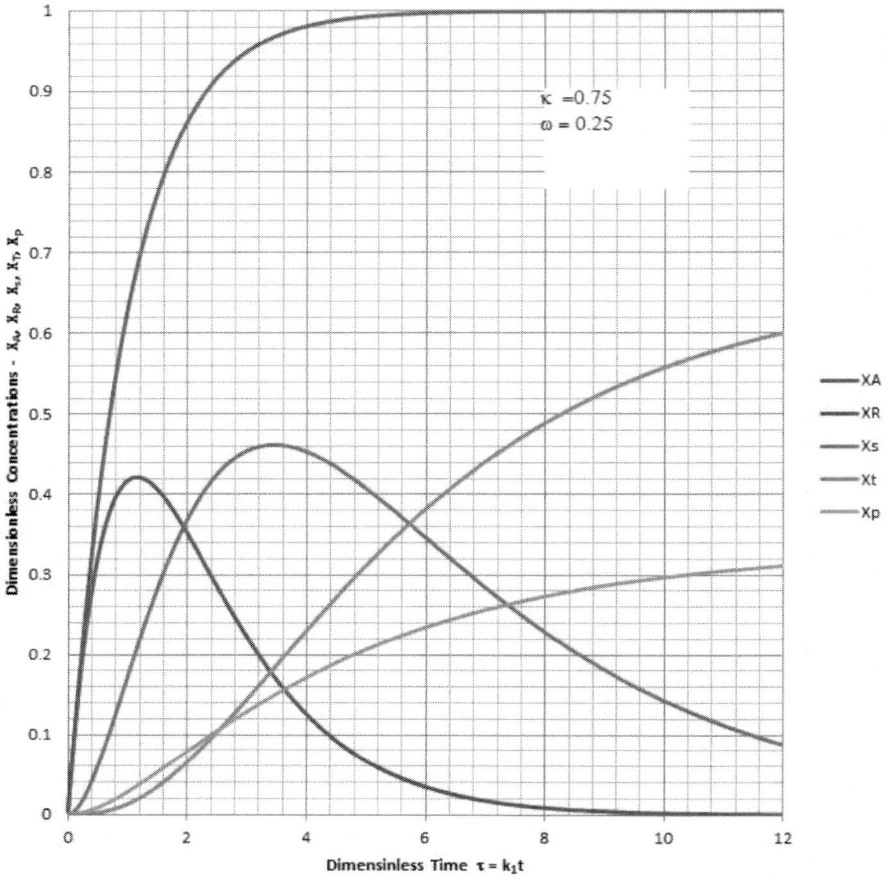

Figure 5.0. Trigylceride (A), Diglyceride (R), Monoglyceride (S), Glycerol (T) and FAME (P) Product Ditribution in Progressive Methanolysis at κ = 0.75 and ω = 0.25.

Computer simulations can be used in order to obtain the torque requirements of a rotor to affect centrifugal separation of liquids with different viscosity and density. A typical CINC centrifugal liquid-liquid separator can be obtained from commercially such as the CINC Processing Equipment, Inc. The CINC Liquid-Liquid Centrifugal Separator utilizes the force generated by rotating an object about a central axis. By spinning two fluids of different densities within a rotating container or rotor the heavier fluid is forced to the wall at the inside of the rotor while the lighter fluid is forced toward the center of the rotor. A cut-away view of such a centrifugal separator may be viewed at the internet webpage http:///www.cincmfg.com/How_our_Centrifuges_Work_s/108.htm.

The theory for separation used currently in the industry is the Stokes settling of oil droplets. For high volume separation such as the biodiesel and glycerol mix from the reactor outflow, a centrifuge such as the one described in this study may be used. Here layers can be expected to form, with one layer that is biodiesel rich and another layer that is glycerol rich. The peripheral layer is glycerol rich and may be collected from a port at the outer centrifugal bowl as shown in Figure 7.0 and the biodiesel-rich layer may be collected from the inner rotor wall that is rotating. There is not much discussion in the literature for the theory of centrifugal separation of *layered flow*.

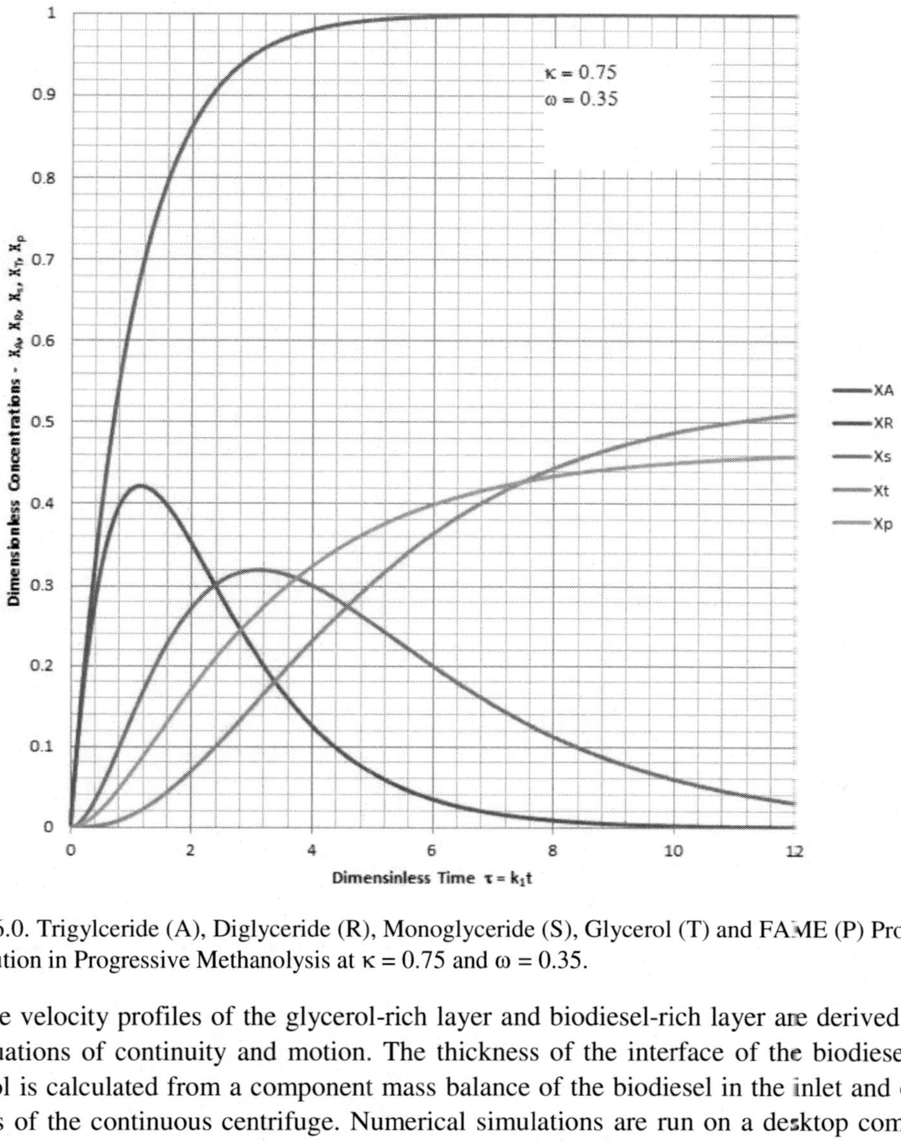

Figure 6.0. Trigylceride (A), Diglyceride (R), Monoglyceride (S), Glycerol (T) and FAME (P) Product Distribution in Progressive Methanolysis at κ = 0.75 and ω = 0.35.

The velocity profiles of the glycerol-rich layer and biodiesel-rich layer are derived from the equations of continuity and motion. The thickness of the interface of the biodiesel and glycerol is calculated from a component mass balance of the biodiesel in the inlet and outlet streams of the continuous centrifuge. Numerical simulations are run on a desktop computer for a given angular speed of rotor, ω (RPM) and *density ratio* of the fluids and viscosities of the fluids. A set of four simultaneous equations and simultaneous unknowns are solved for using the MINVERSE command [44] in Microsoft Excel for Windows 2007. These constants are used to obtain the power draw at the rotor from the torque required. A log-log plot is developed form the simulations for the power draw at the rotor that may be used in the design of such systems.

Shear Flow Theory

Consider a centrifuge with a outer bowl radius of R (m) and an inner rotor radius of κR (m). The inner rotor is allowed to rotate at an angular velocity of ω RPM. The feed has high concentration of biodiesel about 33% mass fraction biodiesel (x_F). It is desired to achieve a separation efficiency of 97.9%. The outlet oil stream is from the inner rotor and the outlet water stream is from the periphery of the bowl. The density ratio of the oil and water is γ. Viscous flows are considered at steady state. Consider a thin shell of fluid with thickness Δr and at a distance r from the center of the centrifuge as shown in Figure 7.0. It is assumed that the momentum transfer is predominantly in the radial direction. The tangential velocity assumes a profile that varies with the distance r from the center of the centrifuge. It is assumed that for high volume feeds two *layers are formed*, i.e., one rich in biodiesel and the second layer rich in glycerol. As the tangential force from the rotor is increased the species with the higher specific gravity will gain more momentum and move to the periphery of the centrifuge. The species with the lower specific gravity will remain in the inner layer close to the rotor. The density of the glycerol was assumed to be "heavy" and was taken as 1260 kg. m^{-3} and the density of the biodiesel was taken as 860 [45] kg.m^3.

For such a pair, the peripheral layer would be glycerol-rich and the inner layer would be biodiesel-rich. Earlier discussions in the literature have been largely on droplet formation of oil and layer formation or "slick" formation is not discussed much. Let the radius of the outer centrifugal bowl that is held stationary be R (m) and that of the inner rotor be κR (m). The inner rotor is allowed to rotate at an angular velocity of ω RPM (revolutions per minute). The water is collected by a port at the periphery of the bowl and the oil is collected through the port in the inner rotor. The feed is introduced from the top of the centrifuge. The feed location has not been optimized in the study. The equation of continuity and motion for v_θ and the equation of motion for shear stress, $\tau_{r\theta}$ can be written from the Appendix in Bird, Stewart and Lightfoot [46] as follows;

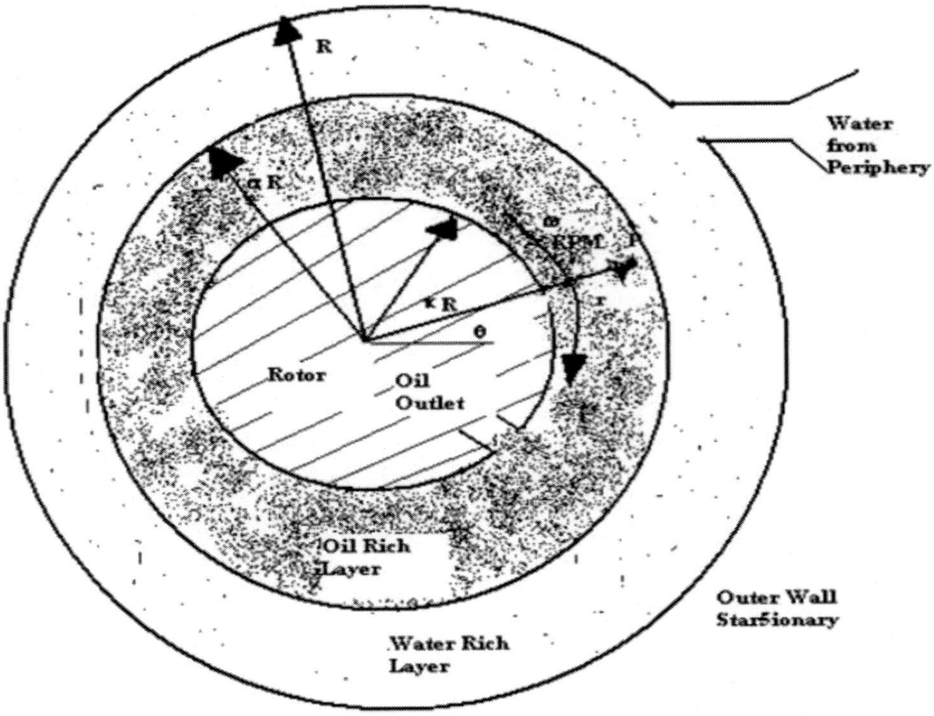

Figure 7.0. Cross-Sectional View of Centrifugal Separator of Oil and Water.

$$-\frac{1}{r^2}\frac{\partial}{\partial r}\left(r^2\tau_{r\theta}\right)=0 \tag{15}$$

Integrating Eq. (15);

$$\tau_{r\theta}=\frac{c_1}{r^2} \tag{16}$$

The Newton's law of viscosity for the shear rate is given by];

$$\tau_{r\theta}=\tau_{\theta r}=-\mu\left(r\frac{\partial}{\partial r}\left(\frac{v_\theta}{r}\right)\right) \tag{17}$$

For the oil rich inner layer (Figure 7.0) combining Eq. (16) and Eq. (17);

$$-\mu_{oil}\frac{\partial}{\partial r}\left(\frac{v_\theta}{r}\right)=\frac{c_1}{r^3} \tag{18}$$

Integrating Eq. (18) twice;

$$\frac{v_\theta}{r}=\frac{c_1}{2\mu_{oil}r^2}+c_2 \tag{19}$$

Eq. (19) is valid for, $\kappa R \leq r \leq \alpha R$.

For the water rich peripheral layer (Figure 7.0), in a similar manner the tangential velocity of the fluid can be written as follows;

$$\frac{v_\theta}{r}=\frac{c_3}{2\mu_{water}r^2}+c_4 \tag{20}$$

Eq. (20) is valid for, $\alpha R \leq r \leq \kappa R$.

The boundary conditions can be seen to be;
at the outer stationary wall,

$$r = R,\ v_\theta = 0 \tag{21}$$

Substituting Eq. (21) in Eq. (20);

$$0=\frac{c_1}{2\mu_{oil}R^2}+c_2 \tag{22}$$

at the inner rotor wall,

$r = \kappa R,\ v_\theta = \omega\kappa R$ (23)

Substituting Eq. (23) in Eq. (20);

$$\frac{v_\theta}{\kappa R} = \frac{c_1}{2\mu_{oil}\kappa^2 R^2} + c_2 \tag{24}$$

at the interface of oil rich and water rich layer,

Interface is assumed to be without any accumulation of forces;

$$\tau_{r\theta}(oil) = \tau_{r\theta}(water)$$

$$\frac{c_1}{\alpha^2 R^2} = \frac{c_2}{\alpha^2 R^2} \tag{25}$$

The velocity across the interface of oil rich and water rich layer is assumed to be continuous;

$$\frac{v_\theta}{\alpha R} = \frac{c_1}{2\mu_{oil}\alpha R^2} + c_2 = \frac{c_3}{2\mu_{water}\alpha^2 R^2} + c_4 \tag{26}$$

In this study, Eqs. (22, 24-26) were used to solve for the integration constants, c_1, c_2, c_3 and c_4 using the MINVERSE function in Microsoft Excel for Windows 2007. The set of equations Eq. (22,24-26) that are needed to obtain the integration constants are given in the matrix form as follows;

$$\begin{pmatrix} 0 & 0 & \frac{1}{2\mu_{water}R^2} & 1 \\ \frac{1}{2\mu_{oil}\kappa^2 R^2} & 1 & 0 & 0 \\ 1 & 0 & -1 & 0 \\ \frac{1}{2\mu_{oil}\alpha^2 R^2} & 1 & -\frac{1}{2\mu_{water}\alpha^2 R^2} & -1 \end{pmatrix} \begin{pmatrix} c_1 \\ c_2 \\ c_3 \\ c_4 \end{pmatrix} = \begin{pmatrix} 0 \\ \omega \\ 0 \\ 0 \end{pmatrix} \tag{27}$$

Eq. (27) is a set of four simultaneous equations and four unknowns. The vector of constants can be obtained as follows;

$$\begin{pmatrix} c_1 \\ c_2 \\ c_3 \\ c_4 \end{pmatrix} = \begin{pmatrix} 0 & 0 & \dfrac{1}{2\mu_{water}R^2} & 1 \\ \dfrac{1}{2\mu_{oil}\kappa^2 R^2} & 1 & 0 & 0 \\ 1 & 0 & -1 & 0 \\ \dfrac{1}{2\mu_{oil}\alpha^2 R^2} & 1 & -\dfrac{1}{2\mu_{water}\alpha^2 R^2} & -1 \end{pmatrix}^{-1} \begin{pmatrix} 0 \\ \omega \\ 0 \\ 0 \end{pmatrix} \quad (28)$$

The layer thickness ratio, α can be estimated as follows. A component balance on the oil in the feed stream, peripheral water stream and inner rotor oil stream would yield;

$$x_F v = (v - v_{rot})x_{per} + v_{rot}x_{rot} \quad (29)$$

or,

$$\frac{v_{rot}}{v} = \frac{(x_F - x_{per})}{(x_{rot} - x_{per})} \quad (30)$$

Let the residence time of the fluid in the continuous centrifuge be θ (hr). Then;

$$v_{rot}\theta = \pi R^2(\alpha^2 - \kappa^2)H \quad (31)$$

and,

$$v\theta = \pi R^2(1 - \kappa^2)H \quad (32)$$

Dividing Eq. (2.80) by Eq. (2.81) and Equating with Eq. (16);

$$\alpha = \sqrt{(1-\kappa^2)\left(\frac{x_F - x_{per}}{x_{rot} - x_{per}}\right) + \kappa^2} \quad (33)$$

The calculations were performed for an rotor speed of 1000 RPM. The separation efficiency is about 97.9%. The values in bold face are obtained by using the MINVERSE command in Microsoft Excel for Windows 2007. The results are the inverse of the matrix as described in Eq. (28). Simulations were repeated for 29 different values of angular speeds of rotor. Each of the torque values were recorded in another column in the spreadsheet. The torque is calculated from the shear stress the rotor wall multiplied with the surface area of the rotor and the moment arm distance, κR, and multiplied with the angular speed ω in RPM.

Table 2.0. Calculations for a Given Set of Oil and Water Viscosities and ω = 10 RPM

Oil Water Separation by Centrifugation				**Feed**	**Outlet (Oil)**	**Outlet (Water)**	
			x_{oil}	0.33	0.99	0.01	
μ_w	0.001	Pa.S	x_w	0.67	0.01	0.99	
μ_{oil}	1000	Pa.S	v_{rotor}	13.3877551			
κ	0.74		v_{per}	27.6122449			
$1-\kappa^2$	0.4524		Separation	0.979591837	0.326531	Ratio	
α	0.83386		Efficiency				
R	5	10000	Gallons	V			
ρ_{oil}	900	$kg.m^3$	Ft	H	0.522028	m	
ρ_{water}	1000	$kg.m^3$					
θ_{res}	1	hr					
v	41	$m^3 \cdot h^{-1}$					
γ	0.9		0	0	0.00002	1	
α	0.83386		3.6523E-05	1	0	0	
			1	0	-1	0	
			2.87636E-05	1	-28.7636	-1	
c1	34.76614		-0.034766137	0.034766137	1	-0.034766137	0
c2	999.9987		1.26976E-06	0.99999873	-3.7E-05	1.26976E-06	1000
c3	34.76614		-0.034766137	0.034766137	-2.7E-07	-0.034766137	0
c4	-0.0007		1.000000695	-6.9532E-07	5.4E-12	6.95323E-07	0
	RPM						
T	ω	α	Torque				
199	1000	0.83386	42.32546536				

The results of these simulations are shown in Figure 8.0 on a log-log plot. The relationship is found to be linear in the log-log plot. For the example run as shown in Table

2.0 the separation efficiency is about 37%. Inorder to achieve more separation more stages need be considered.

The set of simulations were repeated for a higher viscosity of oil, μ_{oil} (5000 Pa.s). The power draw at the rotor is also shown in Figure 8.0 in the log log plot. The increase in power draw corresponding to an increase in viscosity of oil was not high.

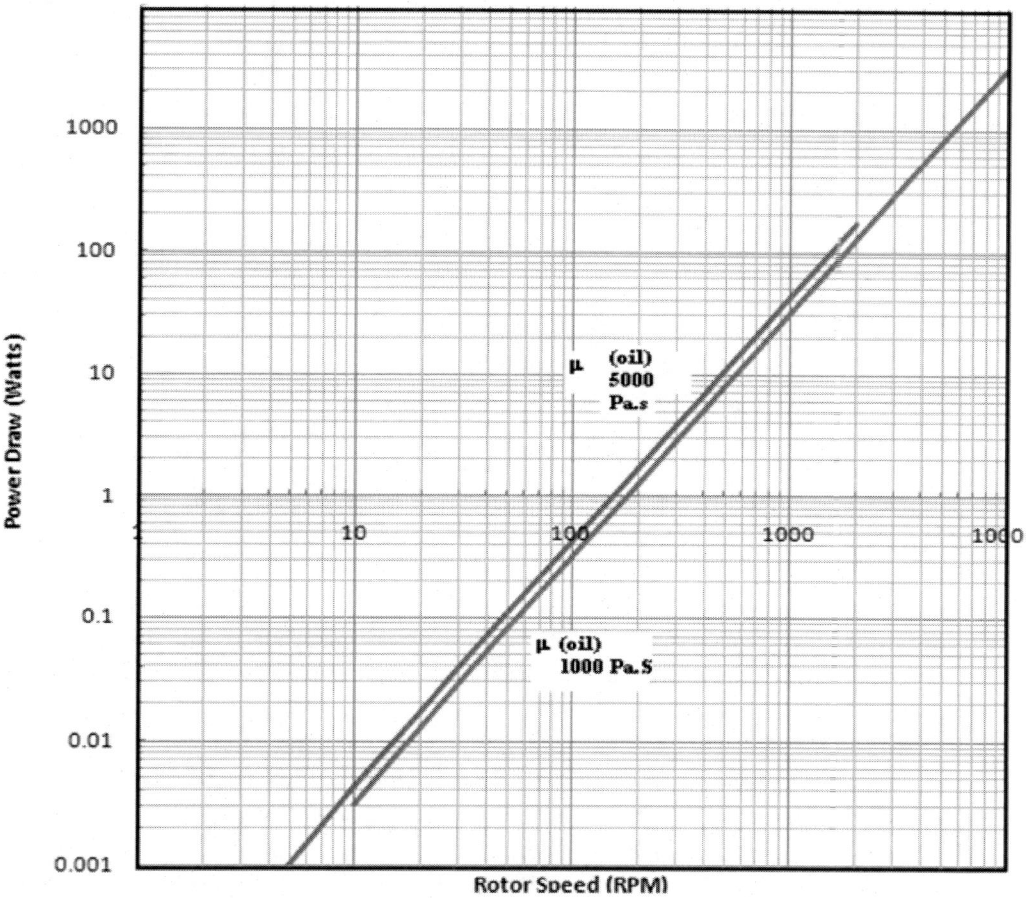

Figure 8.0. Power Draw as a Function of Rotor Speed for κ = 0.74.

CONCLUSION

Advances in microarray analysis, sequencing of genomes may lead to the improvement in cultivability of *J. Curcas* shrub. The ribosome inhibiting protein action and regulation of gene expression studies from transfected clones of the curcin gene from the *J. Curcas* and other studies may increase the yield of the *Jatropha Oil* from the shrub. NGS, Next-Generation Sequencing machines that are rapidly becoming less expensive from different vendors are compared side by side and tabulated. The genominomics of sequencing *J. Curcas* has changed rapidly in recent years and holds promise for higher quality shrubs. The pathway that is more probable during biodiesel formation from triglycerides in *Jatropha Oil* obtained from quantum orbital calculations is of the consecutive-competitive type. The yield of biodiesel over glycerol can be higher for some reaction rate constant ratios and *vice versa* for some reaction rate constants. The results from model solutions are shown in Figures 2.0 – 6.0. The centrifugal separation of glycerol and biodiesel are simulated using Microsoft Excel Spreadsheet. For a given density ratio and viscosity ratio the power needed for the rotor as a function of RPM is obtained from the simulations and found to be log linear as shown in Figure 8.0. Plot was obtained for 29 different values of the agitator RPM and two liquids with different viscosities. The layer formation may become unstable at certain volume fractions of the feed. The velocity profiles for Coutte flow in the centrifuge when layers are formed are obtained. The torque is calculated from the shear stress the rotor wall multiplied with the surface area of the rotor and the moment arm distance, κR, and multiplied with the angular speed ω in RPM. The conversion of *Jatropha Oil* in the reactors may be optimized for minimization of total cost of reactor and separator.

REFERENCES

[1a] http://www.opec.org/opec_web/en/data_graphs/330.htm.

[1] Wetterstand, K. A. DNA Sequencing Costs: Data from the NHGRI Genome Sequencing Program. 2014. http://www.genome.

[2] Sharma, K. R. $1000 Genome is within Reach. *Journal of Pharmaceutical and Biomedical Sciences,* July 2014.

[3] Sharma, K. R. Microarray Analysis:Biochips and Eradication of Disease. New York; Momentum Press. (at press).

[4] Dehgan, B. Comparitive Anatomy of the Petiole and Infrageneric Relationships in the Genus Jatropha (Euphorobiaceae). *Amer. J. Bot.,* 1982; 69:1283-1295.

[5] Fairless, D. Biofuel: The Little Shrub that could – Maybe. *Nature,* 2007; 449:652-655.

[6] Heller, J. Physic Nut J. Curcas L. Promoting the Conservation and Use of Underutilized and Neglected Crops. Institute of Plant Genetic and Crop Plant Research, Gatersleben/International Plant Genetic Resource Institute, Rome, Italy. http://www.ipgri.cgiar.org/Publications/pdf/ 161.pdf.

[7] Abdulla R., Chan E. S., Ravindra P. Biodiesel Production from J. Curcas: A Critical Review. *Critical Reviews in Biotechnology*, 2011; 21(1):53-64.

[8] Jones N., Miller J. H. J. Curcas: A Multipurpose Species for Problematic Sites. *Land Resources*, 1992, 1:12.

[9] Kumar A., Sharma S. An Evaluation of Multipurpose Oil Seed Crop from Industrial Uses (J. Curcas L): A Review. *Ind Crops Prod.*, 2008, 28:1-10.

[10] Rathore V. G., Madras G. Synthesis of Biodiesel from Edible and non-Edible Oils in Supercritical Alcohols and Enzymatic Synthesis in Supercritical Carbon Dioxide. *Fuel,* 2007, 86:2650-2659.

[11] Inekwe U. V., Odey M. O., Gauje B., Dakare A. M., Ugwumma C. D., Adegbe E. S. Fattyacid Composition and Physicochemical Properties of J. Curcas Oils from Edo and Kaduna States of Nigeria and India. *Annals of Biological Research,* 2012, 3 (10):4860-4864.

[12] Shah S., Sharma A., Gupta M. N. Extraction of Oil from J. Curcas L. Seed Kernels by Combination of Ultrasonification and Aqueous Enzymatic Oil Extraction. *Biresource Technology*, 2005; 96:121-123.
[13] http://en.wikipedia.org/wiki/ASTM_D6751.
[14] http://www.marketresearchmedia.com/?p=630.
[15] The Hindu, Chennai, India, February 27th 2006.
[16] http://www.nytimes.com/2006/09/12/business/smallbusiness/12bio.html?fta=y&_r=0.
[17] Podolski W. F., Shmalzer D. K., Conrad V., Lowenhaupt D. E, Winschel R. A., Klunder E. G., Mcllvried H. G., Ramezan M., Stiegel G. J., Srivastava R. D., Winslow J., Loftus P. J., Benson C. E., Wheeldon J. M., Krumpelt M., Lee-Smith F. Energy Resources, Conversion, and Utilization. Perry's Chemical Engineers' Handbook, Eds. Green D. W., Perry R. H. New York. McGraw-Hill, 2008.
[18] Congressional Record. 2005; 51 (4):4544.
[19] Shay E. G. Diesel Fuel from Vegetable Oils: Status and Opportunities. *Biomass and Bioenergy*, 1993; 4:227-242.
[20] Abraham M. A., Nguyen N. Results from the Sandestin Conference: Green Engineering: Defining Principle. *Environmental Progress*, 2005; 22:233-236.
[21] Krawczyk T. Biodiesel – Alternative Fuel makes Inroads but Hurdles Remain. *INFORM,* 1Vol. 7, 1996; 7 (1): 801-829.
[22] Sharma K. R. Power Draw of the Rotor during Centrifugal High Volume Separation of Oil and Water. *Journal of Mechanical Engineering Research,* 2012; 4 (1): 10-16. http://www.academicjournals.org/jmer/ PDF/Pdf2012/January/Sharma.pdf.
[23] Connemann J., Krallmann A., Fischer E. Process for the Continuous Production of Lower Alkyl Esters of Higher Fatty Acids. US Patent 5,354, 878. 1994.
[24] Martin A., Flynn J. E., Lange H. Biodiesel Production Method. US Patent 8,030,505. 2011.
[25] Kulkarni M. G., Dalai A. K. Waste Cooking Oil – An Economic Source for Biodiesel: A Review. *Industrial and Engineering Chemistry*, 2006; 45 (9):2901-2913.
[26] Ma F., Hanna M. A. Biodiesel Production: A Review. *Bioresource Technology,* 1999; 70: 1-15.
[27] Qin X., Zhang J., Shao C., Lin S., Jiang L., Zhang S., Xu Y., Chen F. Isolation and Characterization of a Curcin Promoter From J. Curcas L. and its Regulation of Gene Expression in Transgenic Tobacco Plants. *Plant Mol. Biol. Rep*., 2009, 27: 275-281.
[28] Zhang L., Zhang C., Wu P., Chen Y., Li M., Jiang H., Wu G. Global Analysis of Gene Expression Profiles in Physic Nut (J. Curcas L.) Seedlings Exposed to Salt Stress. *PLOS ONE*, 2014, 9 (5): 1-9.
[29] Sato S., Hirakawa H., Isobe S., Fukai E., Watanabe A., Kato M., Kawashima C., Minami C., Muraki A., Nakkazaki N., Takahashi C., Nakayama S., Kishida Y., Kohara M., Yamada M., Tsuruoka H., Sasamoto S., Tabata S., Aizu T., Toyoda A., Shin-i T., Minakuchi Y., Kohara Y., Fujiyama A., Tsuchimoto S., Kajiyama S., Makigano E., Ohmido N., Shibagaki N., Cartagena J. A., Wada N., Kohinata T., Atefeh A., Yuasa S., Matsunaga S., Fukui K. Sequence Analysis of the Genome of an Oil-Bearning Tree, J. Curcas L. *DNA Research*, 2011, 18:65.
[30] Silva-Junior O. B., Rosado T. B., Laviola B. G., Pappas M. R., Pappas G. J., Grattapaglia D. Genome-wide SNP Discovery from a Pooled Sample of Accessions of the Biofuel Plant J. Curcas based on Whole-Transcritome Illumina Sequencing. IUFRO

Tree Biotechnology Conference 2011: From Genomes to Integration and Delivery, 2011, Arraial d'Ajuda, Bahia, Brazil, June 26th – July 2nd.

[31] Li C., Li L., Fu, Q., Niu, L., Xu, Z. F. Isolation and Functional Characterization of JcFT a Flowering Locus (FT) Homologous Gene from the Biofuel Plant J. Curcas. *BMC Plant Biology*, 2014, 14:125.

[32] Sharma K, R. Fundamentals of Engineering Economics. San Diego. 2012, Cognella Academic.

[33] Bender M. Economic Feasibility Review for Community-Scale Farmer Cooperatives for Biodiesel. *Bioresource Technology*, 1999; 70 (1):81-87.

[34] Weber J. A. The Economic Feasibiloty of Community-Based Biodiesel Plants. Master's Thesis. University of Missouri, Columbia, MO.1993.

[35] Nelson R. G., Shrock M. G. Energetic and Economic Feasibility associated with the Production, Processing and Conversion of Beef Tallow to Diesel Fuel. in Proc. of First Biomass Conference of the Americas: Energy, Environment, Agriculture and Industry. *NREL, National Renewable Energy Laboratory, Golden, CO.,* 1993; 2: 848-862.

[36] Turton R., Bailie R. C., Whiting W. B., Shaeiwitz J. A. Analysis, Synthesis, and Design of Chemical Processes. Upper Saddle River, NJ. Prentive Hall. 1998.

[37] Shiva Prasad B. G., Sharma K. R. Alternative Energy for Energy Sustainability. Annals of Arid Zone, Special Issue, Jodhpur, India. April, 2012.

[38] You Y. D., Shie J. L. Chang C. Y., Huang S. H., Pai C. Y., Yu Y. H., Chang C. H. Economic Cost Analysis of Biodiesel Production: Case in Soybean Oil. *Energy and Fuels*, 2008. 22:182-189.

[39] Darnoko D., Cheryan M. Continous Production of Palm Methyl Esters. *Journal of American Oil Chemists Society*, 2000, 77(12):1269-1272.

[40] Om Tapanes N. C., Aranda D. A. G., Carneiro J. W. M., Antunes O. A. C. Transesterification of J. Curcas Oil Glycerides: Theoretical and Experimental Studies of Biodiesel Reaction. *Fuel*, 2008, 87: 2286-2295.

[41] Levenspiel O. Chemical Reaction Engineering. Hoboke, NJ. 1999. John Wiley & Sons.

[42] Mickley H. S., Sherwood T. K., Reed C. Applied Mathematics in Chemical Engineering. New York. 1957, McGraw Hill.

[43] Jaya, N., and Ethirajulu K. Kinetic Modeling of Transesterification Reaction for Biodiesel Production using Heterogeneous Catalyst. *Int. J Engineering Science and Technology,* 2011. 3(4): 3463-3466.

[44] Sharma, K. R. Continuous Process Dynamics, Stability, Automation and Control. Happauge, NY. Nova Science. 2015.

[45] https://cset.stanford.edu/sites/default/files/Chemistry%20of%20Biodiesel%20Production%20Teacher%20Guide.pdf.

[46] Byrd R. B., Stewart W. E., Lightfoot E. N. Transport Phenomena. Uppersaddle River, NJ. 2007. John Wiley & Sons.

In: Jatropha Curcas
Editor: Gregory Medina

ISBN: 978-1-63483-089-8

Chapter 3

POTENTIAL USES OF *JATROPHA CURCAS*

M. Moniruzzaman*[1]*, Parul Akhtar*[1]*,
Zahira Yaakob*[1] *and A. K. M. Aminul Islam*[2]*
[1]Department of Chemical and Process Engineering,
Faculty of Engineering and Built Environment,
Universiti Kebangsaan Malaysia, Bangi, Selangor DE, Malaysia
[2]Department of Genetics and Plant Breeding, Bangabandhu Sheikh
Mujibur Rahman Agricultural University, Gazipur, Bangladesh

ABSTRACT

The member of eupherbiaceae family, Jatropha is being explored as multi dimensional prospective plant. It has social, agricultural, environmental, industrial, pharmaceutical and energy production potentials. Jatropha seed cake, compost from debris and slurry from biogas plant are good sources of soil nutrition. Protein from seed cake can be utilized as animal feed. As a plant Jatropha can assimilate CO_2 from climate and it can also be used as bio adsorbent of heavy metal from effluents. Protein, lignin and other products of Jatropha can substitute the hazard synthetic chemicals of polymers production. Seed oil have been using in soap, dying and cosmetic industry from long before. Extract from Jatropha leaf, bark posses biological active compounds that have medicinal potentials. Jatropha is more familiar as energy crop. It provides energy in the form of firewood, seed oil as lamp, cooking and engine fuel directly. Biodiesel the chemical conversion of seed oil can be used directly or blend with fossil diesel with modified or conventional engine. However, presence of toxic substances (pherbol ester) renders its utilization as therapeutic agents and animal feed. High viscosity of seed oil and biodiesel is the hurdle to use. More importantly, poor yield in field condition than expectation is the main barrier to be commercialized of Jatropha cultivation. It needs to extensive and systematic screening of available germplasoms and application of biotechnology for the development of high yielding cultivar, agronomic practice and crop management for maximum yield, biochemical engineering for the improvement of biodiesel quality and finally customization of engine according to fuel properties.

* Corresponding author: A. K. M. Aminul Islam. Department of Genetics and Plant Breeding Bangabandhu Sheikh Mujibur Rahman Agricultural University, Gazipur 1706, Bangladesh. E-mail: aminuljkkp@yahoo.com.

Keywords: Physic nut, energy, medicine, cosmetics, antioxidant, biodiesel

1. INTRODUCTION

1.1. Geographical Distribution and Domestication

Jatropha is indigenous to Central America and the northern parts of South America. It is believed that from 16th century onwards Jatropha have been introduced to tropical and sub-tropical regions of Asia and Africa by European seafarers and explorers (Brittaine and Lutaladio, 2010). The ability of Jatropha to grow on different types of soil, more tolerance to both biotic and abiotic stress facilitate to introduction and adaptation in different parts of the world. Figure 1 discloses the geographical distribution of Jatropha.

Throughout the world more than 1,000,000 ha of Jatropha have been propagated. Majority (85%) is in Asian countries i.e. India, China and Myanmar. Among the remaining, 12% in Africa and 2% in Latin America (in Brazil and Mexico). India is on top among the Jatropha cultivator countries. (Edrisi et al., 2015).

1.2. Cultivation

Cultivation is pre preparation of harvesting. It includes climate consideration of the cultivated crop, soli selection and preparation, propagation or plantation, agronomical practice or nursing.

Climate

The tropical and sub tropical regions (30°N and 35°S) are the best places for Jartropha production. It can grow from Zero (0) to 500 meters altitudes from sea level (Figure 1). For seed production of Jatropha plant annual rainfall between 1000 and 1500 mm is optimum. For flowering and fruit setting at least 600 mm rainfall is required.

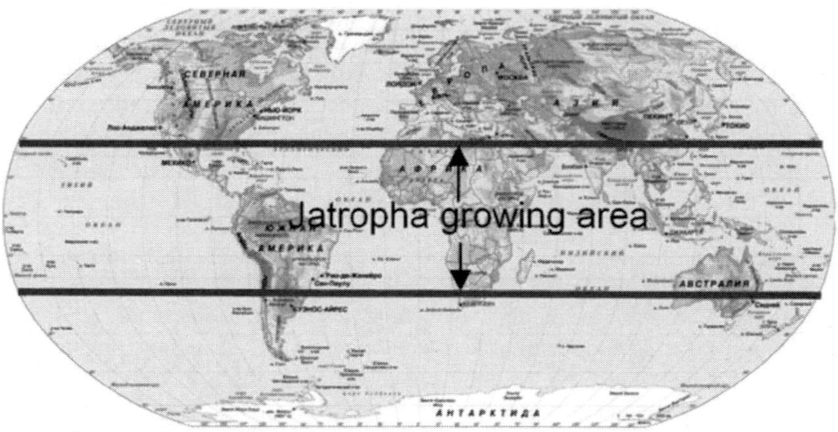

Jongschaap, Corré, Bindraban, and Brandenburg, 2007.

Figure 1. World Jatropha producing areas.

However, the plant can survive with as little as 250 to 300 mm (FACT, 2007). Jatropha growing with 3000 mm of rainfall is more susceptible to fungal attract and poor root growth. High temperature affects negatively on productivity.

The most favorable temperature to grow and production is between 20°C and 28°C (Gour, 2006). Photoperiodism doesn't affect flowering. Flowers and fruits appear at any time of the year if plant is in comfortable condition (Heller, 1996). It is intolerant to frost.

Soils

Well drained and well aerated soil structure is good for Jatropha. According to soil classification based on grain size by USDA; soil of loam, sandy clay loam and silt loam are most suitable for Jatropha cultivation (Gour, 2006). Heavy soil (clay, sandy clay, clay loam, silt clay loam, and silt) can be suitable with low rainfall climate. This condition Jatroph yields more, because these soil contain relatively more nutrient. Jatropha can grow on sand, loamy sand, and sandy loam as they endure draught.

But these soils are barren. It needs to inoculate fertilizers and organic matters for good production. Soil P^H is an important factor for the growth of plants.

Jatropha can grow a wide range of P^H. The optimum P^H range lies between 6 to 8/8.5 (FACT, 2007). Dagar et al. reported that Jatropha can tolerate salinity but it needs to measure the yield performance under this condition (Dagar et al., 2006).

1.3. Production

The productivity of Jatropha greatly varies among individual trees. Theoretical calculation considering model plants, Jongschaap reported 7.8 tonnes per ha seed yield under optimum conditions (Jongschaap et al., 2007). Jatropha plantation on good soil, regular irrigation, proper nursing can yields 5 to 7 tonnes per ha (Achten et al., 2008; FACT, 2007).

On the other hand some other report shows poor yield. Ghokale (2008) and Ouwens et al., (2008) reported the productivity are 1 tonnes per ha and less than 1.25 tonnes per ha respectively (Gokhale, 2008; Ouwens et al., 2008). Some authors reported high fluctuation of productivity 0.1 to 12 tonnes per ha (Heller, 1996; Openshaw, 2000a). This is because of unavailability of exact data, without consideration of agronomic practice, production stage and genetic potential of studied plants.

2. Phyto-Chemistry of Jatropha / Phytochemical Properties

Jatropha curcas is a rich sources of secondary metabolites such as diterpenes, triterpenes, sesquiterpenoids, flavonoids, lignans, neolignans, coumarins, phytosterols and alkaloids. Previous phytochemical studies (Liu et al., 2013; Naengchomnong et al., 1994; Ravindranath et al., 2004b; Wang et al., 2009; Xu et al., 2011) have reported the presence of secondary metabolites in different parts of *J curcas* and these compounds are presented in the next subsequent sections.

2.1. Leaf

Khafagy and co-workers reported two new flavonoidal glycosides (I and II) as well as a new dimer of a triterpene alcohol, stigmasterol and *β*-sitosterol from the leaves in 1977 (Khafagy et al., 1977). In 1990, Chhabra et al., isolated one flavonoids, two glycosides and three sterols namely, apigen, vitexin and isovitexin as well as stigmasterol, *β*-d-sitosterol and *β*-d-glucoside respectively from the leaves of *J. curcas* (Chhabra et al., 1990). Neuwinger also reported sapogenins, 1-triacontanol and a dimer of a triterpene alcohol from the leaves In 1994 (Neuwinger, 1994). Furthermore Staubmann et al., isolated a complex mixture alkaloids such as 5-hydroxypyrrolidin-2-one and pyrimidine-2,4-dione from the same parts of *J. curcas* (Staubmann et al., 1999b). A novel biflavone di-C-glucoside, 6,6″-di-C-β-D-gluco-pyranoside-methylene-(8,8″) -biapigenin was isolated from the leaves of *J. curcas* along with known six compounds such as orientin, vitexin, vicenin II, apigenin 7-O-β-D-neohesperidoside, apigenin 7-O-β-Dgalactoside and apigenin by Abd-Alla et al. (2009). Azzaz and his co-authors reported seven antifungal active flavonoid compounds such as apigenin-7-O-β-D-neohesperidoside, apigenin-7-O-β-D-galactoside, biapigenin-(8,8″)-met-hylene-6,6″-di-C-β-D-glucopyranoside, orientin, apigenin, isovitexin and vitexin from methanolic leaf extract in 2010 (Azzaz et al., 2010).

2.2. Root

The root is an important part of *Jatropha curcas* which can be used for the cure of various diseases such as eczema, gum bleeding, dysentery, gonorrhoea, toothache, ringworm, and scabies (Burkill, 1995; Oliver-Bever, 1986). Naengchomnong and co-workers reported four diterpenes namely curcusones (A-D) reported by Naengchomnong and others from *Jatropha curcus* roots in 1986 (Naengchomnong et al., 1986a).

In the same year they reported another two novel lathyranes diterpenes namely curculathyranes A and B (Naengchomnong et al., 1986b). They also isolated (+)-Jatrophol, (+)-Marmesin, Propacin and Jatrophin from the same parts (Naengchomnong et al., 1994). In 1996 Kong et al., reported thirteen compounds from the roots and were documented as nobiletin, β-sitosterol, taraxerol, caniojane, daucosteroljatropholone A and B, jatropholone B, 5α-stigmastane-3,6-dione, 5-hydroxy-6,7-dimethoxycoumarin, 6-methoxy-7-hydroxycoum-arin, along with one new compound 2S-tetracosanoic acid glyceride-1 and two phenolic compounds such as 3-hydroxy-4-methoxybenzaldehyde, 3-methoxy-4-hydroxybenzoic acid (Kong et al., 1996). Ravindranath et al., reported two pimaranediterpenes from the roots named as 3β-acetoxy-12-methoxy-13-methyl-podcarpa-8,11,13-trien-7one and 3β,12-dihydroxy-13-methyl-podoacrpane-8,10,13-triene (N. Ravindranath et al., 2004a). An antibacterial active compound, 4-Butyl-2-chloro-5-formyl-1H-imidazole reported by Das et al. (2005). A noble diterpenoid lactam jatrophalactam was reported from the roots of *J. curcas* by Wang et al. (2009). In 2011 Aiyelaagbe and co-authors reported eight diterpenes such as curcusone C and D, 15-epi-4Z-jatrogrossidentadion, 4Z-jatrogrossidentadion, 4E-jatrogrossidentadion, multidione, 2-hydroxyisojatrogrossidion and 2-epi-hydroxyiso-jatrogrossidion from the roots (Aiyelaagbe et al., 2011). Chianese and co-workers isolated three new diterpenoids namely, acetoxyjatropholone, curcusone E and spirocurcasone along with eleven (11) known compounds such as (4*E*)-Jatrogrossidentadion, 15-*epi*-(4*E*)-

Jatrogrossidentadion, jatropholone B, curcusones A-B and the remaining compound also reported Aiyelaagbe et al., in the same year from root bark (Chianese et al., 2011). Liu and co-workers collected the roots of *J. curcas* from Xishuangbanna, Yunnan Province. They reported three new diterpene compounds such as jatrophalactone, jatrophalone and jatrophadiketone in 2012 (Liu et al., 2012).

In the same year (Zhang et al., 2012) reported three new diterpenoids namely *epi*-jatrophol, jatrophaldehyde and *epi*-jatrophaldehyde from the root bark along with nine known compounds, which already reported others researcher (Chianese et al., 2011; Naengchomnong et al., 1994; Naengchomnong et al., 1986a; Ravindranath et al., 2004b). In 2013, they also isolated nineteen (19) diterpenes namely curcusones A–J, curcusecons A–E, 4-*epi*-curcusone E,3-dehydroxy-2-*epi*-caniojane, jatrogrossidione and 2-*epi*-jatrogrossidione (Liu et al., 2013).

Yang et al., reported two new sesquiterpenoids, (1*S*, 2*R*)-dihydroxycycloax-4(15)-ene, 14-dehydroxyl daucucarotol, and one new rhamnofalanediterpenoid, 2-hydroxy-3-dehydro-xycaniojane, along with two known compounds, curcusone C-D from the roots (Yang et al., 2013).

Recently Othman has been reported anti-inflammatory active compounds from the roots of *J. curcas* and identified as hexadecanoic acid methyl ester, 9-octadecanoic acid methyl ester and octadecanoic acid by GC-MS (Othman et al., 2015).

2.3. Stem

Mitra and co-workers reported sesquiterpenoids and triterpene compounds such as β-amyrin, taraxerol and β-sitosterol from the stem bark of *J. curcas* in 1970 (Mitra et al., 1970).

In 2004 Ravindranath and his co-workers reported three deoxypreussomerins, namely palmarumycin JC1-JC2 and palmarumycin CP1 from the stems of *J. curcas* (Ravindranath et al., 2004a).

2.4. Seed or Seed Oil

In 1970 Mitra and co-workers isolated β-D- glucoside of β –sitosterol, dulcitol and sucrose from the seed kernel (Mitra et al., 1970).

Hirota et al. reported atumour promoter phorbol esters named 12-deoxy-16-hydroxyphorbol from seed oil (Hirota et al., 1988). Six phorbol ester separated namely Jatropha factors C1–C6 from seed oil of *Jatropha curcus* by Haas et al. (2002).

A ribosome deactivating protein named curcin was reported by Juan et al. (2002). Jing and co-workers reported Jatropherol 1 deoxypreussomerinditerpenes from the seed (Jing et al., 2005). *Jatropha curcas* seeds are very of toxic and allergenic. In 2009, Maciel et al., reported one new allergenic 2S albumin, designated Jat c 1 from the seeds and suggested albumin caused allergy (Maciel et al., 2009). A ribosome-inactivating protein, Curcin isolated many researcher from the seeds of *J. curcas* (Lin et al., 2010; Luo et al., 2006).

Yao et al. (2012) reported one new asymmetric diamide, curcamide along with seven known compounds such as isoamericanin, isoprincepin, caffeoylaldehyde, isoferulaldehyde, glycerol monooleate, syringaldehyde and β-ethyl-D-glucopyranoside from the seed cake (Yao et al., 2012).

Recently Li and co-workers has been reported two new ligninsjatrophasin C and D together with eight known compounds such as jatrophasin A, (±)-3,3′-bisdemethyl-pinoresinol, *β*-sitosterol, daucosterol, isoamericanol A, isoprincepin, 7′-epi-sesamin-dicatechol, and americanol A (Li et al., 2014).

2.5. Latex

Latex is a very important part of *Jatropha curcas* and various researchers used this part to test the pharmacological activity. A few researchers were able to isolate secondary metabolites from this part. In 1991, Nath and Duta extracted a proteolytic enzyme, curcain from the latex of *J. curcas* (Nath and Dutta, 1991). A novel cyclic octapeptide named curcacycline A was reported from the latex by (Van den Berg et al., 1995), whereas Auvin and co-workers isolated another cyclic nonapeptide Curcacycline B from the same part in 1997 (Auvin et al., 1997). A cyclic peptide, Jatrophidin I isolated from the latex of *Jatropha curcas* reported by (Altei et al., 2014).

2.6. Aerial

A rare dinorditerpene compound named heudolotione was isolated by Ravindranath et al., from aerial parts of *J. curcas* (Ravindranath et al., 2003). In 2004, they also reported twenty compounds from this parts and first four compounds were new lathyrane and podocarpane type diterpenoidsnamed 15-*O*-acetyl-15-epi-(4*E*)-jatrogrossidentadione, 3*β*-acetoxy-12-methoxy-13-methyl-podcarpa-8,11, (14*E*)-14-*O*-acetyl-5,6-epijatrogrossidentadione, 13-trien-7one and 3β,12-dihydroxy-13-methyl-podoacrpane-8,10,13-triene.

The remaining sixteen compounds such as tetradecyl (E)-ferulate, 3-O-(Z)-coumaroyloleanolic acid, heudelotinone, epi-isojatrogrossidione, 2a-hydroxy-epiisojatrogro-ssidione, 2-methoxy-antraquinone, scopaletin, tomentin, curcasones A—D, Jatropholones A and B, Jatrophol and 15-epi-(4E)-jatrogrossidentadion (Ravindranath et al., 2004b).

Pertino et al., reported two phorbol esters named jatropholones A and B from *J. curcas* (Pertino et al., 2007). In 2011, Xu and co-authors reported one diterpene, Jatrophodione A alongwith nine known compounds, caniojane, jatropholone A and B, jatrogrossidione, 2-epijatrogrossidione, heudelotinone, gossweilone, (3*α*)-3-hydroxy-*ent*-pimara-8(14),15-dien-12-one and 12-hydroxy-13-methylpodocarpa-8,11,13-trien-3-one from the aerial parts of *J. curcas* (Xu et al., 2011).

3. Pharmacological or Biological Properties

The medicinal plant *Jatropha curcas* has been widely used as a folk drug in various countries. *J. curcas* contain various secondary metabolites with medicinal importance and it has been used to treat various diseases such as arthritis, gout, jaundice, wound, inflammation, anti-diabetic, antitumor etc. However, the nature of crude extract or secondary metabolites involved in these diseases has not been well recognized.

Hence, this study accompanied to explore the various biological activities of *J. curcas* and to identify the active compounds involved. Numerous biological activities of *Jatropha curcas* has been described in the subsequent sections.

3.1. Antibacterial and Antifungal Activity

Jatropha curcas is toxic plants and their crude extracts have significant anti-bacterial and anti-fungal properties. Researchers have been studying its antimicrobial activity from different parts using various methods from last three decades. Gupta and co-workers studied the antimicrobial properties of seed and leaf extract as well as oil seed. They found the inhibition ranging from 7 to 22 mm and 10 to 20 mm for antibacterial and antifungal activity, respectively and seed extract presented important activity against the tested gram-positive bacteria among all the extracts. Seed extract also exhibited weighty activity against *Mucor* and *Tilletia* fungus (Gupta et al., 2010).

Rachana et al., presented that the effectiveness of *J. curcas* fruit against pathogenic microorganisms such as *P aeruginosa*, *S. aureus* and *E. coli.* They observed maximum inhibition zone of 25.5mm for methanolic extracts of Seed Kernel (Rachana et al., 2012). Seeds and leaf extracts have molluscidal, insecticidal and fungicidal properties. Rahman et al., observed higher antifungal activity for the seed and pulp extracts than the fruit extract. They found extracts inhibited the mycelium growth of *Colletotric hummusae* and *C. gloeosporioides* that causes anthracnose disease in bananas and respectively (Rahman et al., 2013). Auwal et al., reported that the extract of root bark showed effective antibacterial activity to treat the anthrax, reproductive tract infection, skin contagion and several stomach sicknesses (Auwal et al., 2013). Recently Dada et al., used an aqueous extract of *J. curcas* to treat the coliform caused diseases (Dada et al., 2014).

Setha and co-workers tested the antibacterial activity of different crude leaves extracts against histamine forming bacteria, namely *Klebsiella pneumoniae* and *Clostridium perfringens* using a maceration method and observed ethyl acetate and methanol methanol extract showed lowest MIC (0.10% w/v) against *C. perfringens* (Setha et al., 2014). Cordova-Albores and co-authors reported the antifungal activity of seed oil to control the gladiolus fungi, *Fusarium oxysporum* f. sp. *Gladioli* and showed this seed oildose of 2.5 mg mL^{-1} inhibited the growth rate to about 0.77 cm day^{-1} (Cordova-Albores et al., 2014).

Numerous researchers tested the various crude extract such as hexane, chloroform, ethyl acetate, dichloromethane, acetone, methanol and ethanol, but most of them observed alkyl alcohol showed the maximum inhibition zone against the bacteria and fungi. Aiyelaagbe observed the root of methanol extract showed broad spectrum activity against *Gardnerella vaginalis*, *Neisseria gonorrhea*, *E. coli*, *S. aureus*, *Klebsiella aerogenes*, *Proteus mirabills*, *P. aeruginosa* except *Candida albicana* (Aiyelaagbe et al., 2007).

Oyi et al., reported effective latex of methanol extract against *Bacillus subtilis, Escherichia coli, Pseudomonas aeruginosa, Stapylococcus aureus, Streptocococcus pyogenes, Candida albicans* and *Trychophyton* sp. (Oyi et al., 2007). Akinpelu and co-workers reported that the crude methanolic leaf extract exhibited antibacterial activity on *Escherichia coli, Stapylococcus aureus* and *Pseudomonas aeruginosa* (Akinpelu et al., 2009).

Kalimuthu et al., has been reported the methanol extract of leaves callus against the *S. aureus* and *Pseudomonas sp* with the inhibition ranging between 20 and 23 mm respectively

(Kalimuthu et al., 2010). Sriprang et al. (2010) showed seed cake of methanol extract against *Stapylococcus aureus, Bacillus cereus, Bacillus megaterium, Escherichia coli, Pseudomonas aeruginosa* and *Salmonella typhi* (Sriprang et al., 2010). Villaseñor and Cariño reported bacterial activity on S. *aureus, B. subtilis, M. phlei, C. albicans* and *Trychphyton mentagropytes* (Villaseñor and Cariño, 2011). The crude methanol extracts not only have antibacterial properties, but also showed antifungal activities reported by (Sharma et al., 2012). Obasi reported significant methanol extract of stem bark antifungal activity among the others fraction or extracts against *C. albicans, C. oryzae, A. niger* and *Saccharromyces cerevisiae* and bacteria *B. subtilis, S. aureus, E. coli.* MIC ranging between 0.625 and 20 μg ml^{-1} (Obasi et al., 2011). Daniyan et al., reported seed of methanol extract against *C. albicans, E. coli, S. typhi*, and *S. aureus* and MIC ranges between 10-25 mg ml^{-1} (Daniyan et al., 2011). Namuli et al. (2011) reported methanol antibacterial activity against *Bacillus subtilis, Stapylococcus aureus, Escherichia coli, Pseudomonas aeruginosa, K. pneumonia* and inhibition zone ranging from 8.0 to 17.7 mm (Namuli et al., 2011). The methanol extract of latex contains saponins and tannins, which reported by (Suhaili et al., 2011) through the phytochemical screening. They also found the antibacterial activity of methanolic latex extract with MIC values ranging from 0.39 mg/ml for methicillin sensitive *S. aureus* and 6.25 mg/ml for *S. marcescens*. The crude ethanolic leaf extract revealed effective antibacterial properties against *P. aeruginosa, E. coli, P. fluorescens* and *S. aureus* reported by Sharma et al. (2010). In 2010, Saetae and Suntornsuk presented the antifungal activity of the ethanol extract against *F. oxysporum, F. semitectum, P. aphanidermatum, L. theobromae, C. lunata, C. capsici*, and *C. gloeosporioides* (Saetae and Suntornsuk, 2010). Arekemase and co-workers tested the antimicrobial activity of hexane, ethanol, water and latex extracts and suggested that ethanolic latex extract to treat the sexually transferred infections (Arekemase et al., 2011). Ekundayo et al., tested the ethanolic leaf and bark extracts of *Jatropha curcas* against *B. subtilis, E. coli, P. vulgaris, S. aureus, S. typhi, P. aeruginosa* and *S. dysenteriae* and observed the inhibition zone from 30.6 to 38.5 mm and MIC ranging from 2.2 to 1 0.0 mm (Ekundayo et al., 2011). In 2014, Kumari observed crude methanol stem extract exhibited highest inhibition zone (13.3±0.04mm) against *Bacillus subtilis* (Kumari et al., 2014).

Jatropha curcas contain many biologically active secondary metabolites, but only few of researcher tested these metabolites anti-bacterial and fungal activity. Wei and co-workers studied the antifungal activity of curcin. It inhibited the hyphal growth and spore formation at concentration of 5μg/mL and observed at dose of 50μg/mL, no spore formed and inhibitory rate of *Pestalotia funereal* on spore formation was 83.8%. They also suggested that the inhibitory effect of curcin might be related to protein disappear of treated fungi (Qin Wei et al., 2004). They also isolated another biological active fungicide, *β*-1,3-glucanase from *J. curcas* and confirmed its activity against *Rhizoctonia solani* and *Gibberelle zeae* (Wei et al., 2005). Ravindranath et al., tested the antibacterial activity of palmarumycins against *S. aureus* with inhibition zone ranges between 10-13 mm (Ravindranath et al., 2004a). Sutthivaiyakit et al., reported antituberculosis activity of caniojane against *Mycobacterium tuberculosis* H37Ra with MIC 25 μg ml^{-1} (Sutthivaiyakit et al., 2009). Azzaz and his co-authors reported seven antifungal active flavonoid compounds such as biapigenin-(8,8″)-methylene-6,6″-di-C-β-D-glucopyranoside, apigenin-7-O-β-D-neohesperidoside, apigenin-7-O-β-D-galactoside, orientin, apigenin, isovitexin and vitexin from crude methanol leaf extract in 2010.

These compounds showed moderate antifungal activity while crude extracts completely inhibit the fungus growth (Azzaz et al., 2010). A cationic antimicrobial peptide (KVFLGLK,

JCpep7) fruitfully identified (Xiao et al., 2011) by the novel cell membrane affinity chromatography method and obtained the MIC ranging from 24 to 64 µg/ml against *S. typhimurium*, *S. dysenteriae*, *P. aeruginosa*, *S. aureus*, *B. subtilis*, and *S. pneumonia.* Phorbol ester is the main biological active compound for *J. curcas.* In 2012, Devappa and co-workers isolated phorbol ester-rich fraction from *J. curcas* oil and observed its significant antibacterial activity against *B. subtilis*, *P. putida*, *Proteus mirabilis*, *S. aureus*, *S. pyogenes* and *E. coli.* MIC ranges 12 and 19 mm and MBC values between 215 and 537 µg ml^{-1}. As well as antifungal activity against *A. niger*, *A. flavus*, *Botrytis cinerea*, *F. oxysporium*, *F. moniliforme*, *Curvularia lunata* and *P. Notatum* (Rakshit et al., 2012). Yao and co-workers isolated a new asymmetric diamidecurcamide from the ethanol extract of the seed cake of *Jatropha curcas* and tested its bacterial activity on *S. aureus*, *B. Subtilise* and *E. coli* and found MIC ranges of 0.25, 0.125 0.5 mg/ml, respectively (Yao et al., 2012).

3.2. Anti-Inflammatory or Analgesic Effects

Many researchers have been extracted anti-inflammatory active extract from different parts of *Jatropha curcas.* In 1997, Staubmann isolated anti-inflammatory active secondary metabolites with the stoichiometries formulas $C_4H_4N_2O_2$ and $C_4H_7NO_2$ from the leaves of ethyl acetate extracts and confirmed by using carrageen an induced rat-paw edema (Staubmann et al., 1997). Uche and Aprioku determined the analgesic and anti-inflammatory effects of leaves of crude methanol extract by acetic acid and egg albumin in mice and Wister albino rats respectively and by egg albumin and induced inflammation in rats. They observed extract doses from10-80mg/kg and compared with 0.5mg/kg dose of standard piroxicam, but above 20mg/kg doses showed higher activity than the standard. Furthermore, crude extract obtained comparable analgesic effects with standard paracetamol (100mg/kg) (Uche and Aprioku, 2008).

Sangeetha et al., also reported the same activity for aqueous leaves and bark extract. They observed maximum analgesic effect for 300mg/kg dose at 60 min with standard aspirin (50mg/kg). They reported effective inflammatory effect for 300mg/kg doses of aqueous leaves extract at one day with 50mg/kg dose of standard Ibuprofen (Sangeetha et al., 2010). The root of *J. curcas* has been used as a pain killer.

Mujumdar and Misar used root of methanol extract and exhibited its significant anti-inflammatory properties in acute carrageenan-induced rat paw edema (Mujumdar and Misar, 2004). Nayak and Patel also reported same test for the alcoholic extract of leaves, stem and roots of *J. curcas* in 2010. They observed alcoholic extracts of three parts exhibited notable anti-inflammatory activity (Nayak and Patel, 2010a).

In another study, they used these three parts and observed alcoholic extracts of stem and root exhibited significant analgesic activity by reducing the yeast-induced pyrexia compared to the references drugs pantazocine and paracetamol (Nayak and Patel, 2010b). In 2011, Oskoueian et al., analysed the crude extracts of leaf, stem bark, root and latex to their inhibitory effects on nitric oxide synthase in macrophages RAW 264.7 cells. They suggested that all the crude extracts of different parts of *J. curcas* were strong iNOS inhibitor and attributed to their anti-inflammatory effects (Oskoueian et al., 2011a).

Olukunle et al., also investigated the analgesic and anti-inflammatory action for the leaves of aqueous extract and suggested that the leaf extract may be useful for the treatment of painful inflammatory conditions (Olukunle et al., 2011). Omeh and Ezeja estimated the analgesic properties of crude methaolic leaves extract. Aacetylsalicylic acid used as standard drug in all the models.

They observed crude methanol extract demonstrated vital analgesic activity (Omeh and Ezeja, 2013). Recently Ameyaw et al., studied the analgesic activity of crude ethanolic extract in acute and chronic muscular pain by vaccinating carrageenan (3%) into the gastrocnemius muscles of rats. Morphine used as a standard analgesic drug. They observed crude extract and morphine enlarged the grip strength in the acute muscular hyperalgesia trial and both of the drugs meaningfully reduced chronic muscle hyperalgesia (Ameyaw et al., 2014). Recently Asuk and co-workers reported that the leaf, stem bark and root of ethanol-methanol (1:1) extracts possess antioxidant activities, but their response to the liver tissue enzyme systems of SOD and CAT vary in either to playing a compensatory role or boosting the activities of the antioxidant enzymes (Asuk et al., 2015). In 2015, Othman et al., identified three compounds such ashexadecanoic acid methyl ester, octadecanoic acid methyl ester and octadecanoic acid from root extract and suggested that these compounds could be responsible for the anti-inflammatory drug (Othman et al., 2015).

3.3. Anticancer or Cytotoxic Activity

Cancer is a spirited source of global death with accumulative illness and mortality. It can spread out any part of the body. Now a day's researcher has been trying to develop new drugs from this herbal plant metabolites and tested its drug activity. Balaji et al., used the crude methanolic leaves extract on melanoma cells. They observed this extract effectively inhibited the metastatic colony formation in B16F10 melanoma cells of C57BL/6 mice lungs (Balaji et al., 2009a). Oskoueian and co-workers studied the methanol extract from leaf, root and stem bark of *J. curcas* on HT-29 cell line and observed the methanolic root extract had anticancer therapeutic property against human colon adeno carcinoma (HT-29) cell line, but its cytotoxic effect on human hepatocyte (Chang cell) was high (Oskoueian et al., 2011b). The different parts of this plant have different biological activity because of various types and levels of secondary metabolites. Anticancer active diterpenes isolated from *J. curcas* plant has been described in numerous papers. Stripe and co-workers isolated anticancer active proteins, curcins from *J. curcas* (Stripe et al., 1976). Muangman and co-workers isolated curcusone B from *J. curcas* and tested the antimetastatic activity against the four human cancer cell lines. At non-cytotoxic doses, it showed significant activity against the metastatic cancers (Muangman et al., 2005). Thippornwong and Tohtong also isolated curcusone B from *J. curcas* and revealed its anti-invasive effect on KKU-100 cells (Thippornwong and Tohtong, 2006). Luo and co-workers reported antitumor activity of curcin in *E. coli* strain M15. At the concentration of 50 μg ml^{-1} the curcin inhibited the growth of tumour cells of cellule pulmonary cancer and gastric cancer in a concentration-dependant mode (Luo et al., 2007). Pertino et al., studied the gastro-protective property for both the jatropholone A and jatropholone B in an HCl/EtOH-induced gastric lesion model in mice. They observed jatropholone A decreased the lesion by 54% at dose of 100 mg/kg, while jatropholone B reduced the lesions by 83-91% at all the doses (Pertino et al., 2007).

Theoduloz et al., described the anti-proliferative property of jatropholone B against lung fibroblasts, gastric adenocarcinoma, lung, bladder and leukaemia cell. While Jatroph one showed significant anticancer activity (Theoduloz et al., 2009). The cytotoxic activity of caniojane against African green monkey kidney fibroblasts (IC_{50}12.9 μg ml^{-1}) and anti-tuberculosis effects against *M. tuberculosis* H37Ra reported by (Sutthivaiyakit et al., 2009). In contrast, Wang and co-authors described insignificant inhibitory property of Jatrophalactam against human lung and colon cancer and epidermal squamous cell carcinoma A431 (Wang et al., 2009).

Nakazibwe determined the antiproliferative action of curcusone B against the chronic myelogeneous leukaemia (K562) and lung carcinoma (H1299) cell. They suggested that the curcusone B could be a potential an anticancer agent (Nakazibwe et al., 2011).

In 2012 Liu and co-workers tested cytotoxic activity of three compound's namely Jatrophalactone, Jatrophalone, Jatrophadiketone and observed only Jatrophalactone showed effective activity against the human myeloid leukaemia HL-60, hepatocellular carcinoma SMMC-7721, lung cancer A-549, breast cancer MCF-7, and colon cancer SW480 cell lines (Liu et al., 2012). *Jatropha curcas* is the rich source of phorbol esters (PEs), which promote tumours (Goel et al., 2007). Oskoueian et al., also isolated PEs from *J. curcas* and evaluated its cytotoxic effects and mode of actions on breast and cervical cancer cell. It submissive cell proliferation. The isolated PEs stimulated the PKC-δ and reduced the proto-oncogenes. As a results caspase-3 protein activated and cell death happened in breast (MCF-7) and cervical (HeLa) cell. They also suggested that this PEs alternative chemotherapeutic drugs for cancer therapy (Oskoueian et al., 2012).

3.4. Antidiabetic Activity

Diabetes is a global problem and researchers have been doing their work entire the world for searching natural drugs that can effectively manage this problem. Traditionally the tea of *J. curcas* leaves (Jaiswal, 2010), decoctions of boiled leaves or fruit, burnt into ashes (Gbolade, 2009), aqueous bark extract mixed with milk (Jayakumar et al., 2010) are used to control diabetes.

However, the scientific information available regarding human models is very scarce and research is needed to cover this aspect in the near future. Mishra et al. (2010) reported anti-hypoglycaemic activity of 50% ethanolic leaf extract of *J. curcas* by oral administration in allaxon-induced diabetic rats. They observed this extract reduced the glucose level in treated rats was 219.5-116.5 and 237-98.8 for the doses of 250 and 500 mg kg^{-1}, respectively, as compared to standard glibenclamide 232-94.5 at 600 μg kg^{-1} dose after treatment. This extract also significantly reduced the cholesterol and triglyceride levels in the rats (Mishra et al., 2010).

Aladodo and co-workers followed the same procedure for aqueous root extract of *J. curcas* and suggested that the root extract possesses hypoglycaemic activity and it improve the deviations in the blood parameters (Aladodo et al., 2013). Recently Johnson et al., evaluated the antidiabetic effect of the methanolic leaf extract in Wistar strain rat. The animals treated with glibenclamide, 150 and 250 mg/Kg body weight of *J. curcas* doses decreased blood sugar level. The extract significantly reduced total bilirubin and liver biomarker enzymes (AST, ALT and ALP) (Johnson et al., 2014).

Farag and co-authors reported the hypoglycaemic effects various extracts of aerial parts. They observed ethanol extract showed significant hypoglycaemic and antidiabetic activities at the dose of 400 mg/kg. Moreover, the drug found to be safe up to doses of 5 g/kg (Farag et al., 2014). In the running year, Asuk and his co-workers discovered the ant diabetic effects of ethanol-methanol (1:1) ratio of leaf, stem bark and root extract of *J. curcas* on the blood and liver tissue of diabetic albino Wistar rats. They suggested that the extracts possess anti-hyperglycemic and antioxidant activities, but their response to the liver tissue enzyme systems of SOD and CAT vary in either to playing a compensatory role or boosting the activities of the antioxidant enzymes. The extract also retreating the oxidative damage tissue caused by diabetes to prevent lipid peroxidation (Asuk et al., 2015).

3.5. Antioxidant Activity

The extract of leaves, latex, stems, nodes and roots of *J. curcas* presented antioxidant activity. El Diwani et al., studied the antioxidant property of ethanolic extracts of various parts of *J. curcas* using the DPPH radical assay. They revealed that the roots and stem were the strongest antioxidant activity than the leaves and nodes. The crude roots extract showed higher free radical scavenging activity and maximum inhibition of 0.521 mg ml^{-1} (El-Diwani et al., 2009). *J. curcas* is a potential source of natural antioxidants, which contain many polyphenolic compounds. Igbinosa and co-workers assessed the polyphenolic contents, antioxidant activity of stem bark extracts. They showed methanol and ethanol extracts contain higher phenolic compounds, which have higher antioxidant activity than the aqueous extracts (Osamuyimen O Igbinosa et al., 2011). In the same year, Oskoueian reported that the latex and leaf extracts of *J. curcas* showed the maximum antioxidant activity as compared to the root and stem bark, which showed moderate (IC_{50} = 57.9 μg ml^{-1}) and weak activities (IC_{50} = > 200 μg ml^{-1}), respectively (Oskoueian et al., 2011a). They also reported methanolic extract exhibited higher antioxidant activities than hot water extract (Oskoueian et al., 2011b). *Jatropha curcas* is a good source of bioactive peptides. Gallegos-Tintoré et al., separated six peptide fractions by gel filtration chromatographic method through the 50 min *J. curcas* protein hydrolysate. They observed 31.7% of the degree of hydrolysis and lower-molecular-weight peptidicfractions showed maximum antioxidant and chelating activities (Gallegos-Tintoré et al., 2011). Oloyede and co-authors assessed the radical scavenging activity of aqueous leaves extract. They observed IC_{50} = 21.2 and 15.67 mg/mL for DPPH and hydrogen peroxide as compared to Butylatedhydroxyanisole (BHA) and directed that the activity was less than the standard (Oloyede et al., 2012). Oilseed cake of *J. curcas* possess both antioxidant and hypolipidemic activities and lowered serum triglycerides, cholesterol and glucose level from blood reported by (Boudjeko et al., 2013). Recently, Li et al., determined the antioxidative active compounds such as jatrophasin C, jatrophasin D from the dried seeds of *Jatropha curcas* (Li et al., 2014).

3.6. Antiviral or Anti-HIV Activity

Muanza and co-authors used a crude methanol extract of *J curcas* in the AIDS-antiviral screening and observed this extract showed a moderate cycloprotective activity against HIV in cultured human lymphoblastoid CEM-SS cells (Muanza et al., 1995). Matsuse and co-workers investigated the anti-HIV effects for water and methanolic extracts from *J. curcas*.

They observed an aqueous extract strongly inhibited the cultured cells, namely HIV protease enzymes and HIV-reverse transcriptase with low cytotoxicity ($CC_{50} > 1000$) (Matsuse, Lim, Hattori, Correa, and Gupta, 1998). Antiretroviral therapies have been used to decrease activity of virus in AIDS patients; the perseverance of dormant virus inhibited thee laminations of virus. Prostration and 12-deoxyphorbol-13-phenylacetate (DPP) stimulate the latent virus, which practically synthesized Paul A. Wender et al., from natural sources (*J. curcas*). This synthesized materials could be used to treat the HIV (Wender et al., 2008). In 2012, Dahake et al., reported drug-resistant HIV and evaluated the leaf extract of *J. curcas*. They found itsanti-viral propertyon drug-resistant HIV (Dahake et al., 2012).

Recently Influenza is a serious respiratory illness and its vaccine can inhibit influenza virus infection. Recently, Patil and co-workers obtained natural antiviral agents from *J. curcas* against influenza A (H1N1) virus. They also used aqueous and methanol leaf extracts to inhibit hemagglutinin protein of influenza virus. They found aqueous and methanol extracts to be nontoxic to Madindar by canine kidney cells below the concentration of 15.57 and 33.62 mg/ml respectively. They also confirmed that the extract of *J. curcas* had a direct effect on the process of virus adsorption leading to its inhibition (Patil et al., 2013).

3.7. Wound Healing Activity

The traditional use of different portions of *Jatropha curcas* for wound healing is recognised in various areas of the world (Balangcod and Balangcod, 2011; Neuwinger, 1996; Osoniyi and Onajobi, 2003).

However, application of scientific investigation on the human body is deficient. Salas and co-authors reported the cicatrization activity of latex onclinical wound in mice skin. They observed wound healing effect only in male Balb/c mice at a single dose by 10, 50 or 100% latex and multiple doses with dilutions between 5 and 10% (Salas et al., 1994).

In 1997, Villegas and co-authors reported significant wound-healing activity of *J curcas* extracts on cell proliferation and exhibited mutagenic activity (Villegas et al., 1997). Shetty et al., evaluated the same activity for bark extract in rats and observed this activity accelerated by developing the skin and granulation tissue, breaking strength, wound shrinkage, weight of dry granulation tissue and levels of hydroxyproline (Shetty et al., 2006).

Esimone et al. (2008) tested the cicatrizant activity by the formation of herbal ointment containing methanol leaf extract from *J. curcas*. They used this ointment to complete wound closure at intervals of 3 days. Gentamicin ointment (1%) used as a standard.

The crude methanol extract formulated ointment (1.5 g/10g ointment) showed higher rates of wound healing property and reduced the epithelialisation time to 14.8 days (Esimone et al., 2008). Sachdeva and co-workers also formulated the ointment from stem barks and followed excision and incision models. They observed wound contraction, hydroxyproline content, tensile strength and histopathological parameters.

All the parameters showed significant ($P < 0.01$) changes compared to control group (Sachdeva et al., 2011).

3.8. Anti-Tumor Activity

Many plants contain ribosome-inactivating proteins (RIP), which are capable of inactivating ribosome by the depurinating eukaryotic ribosomal RNA. There are two types of RIPs in plants, which are single chain proteins and double chain RIPs. Since RIP is a cell-killing agent, the RIPs have been shown to selectively kill tumour cells. Curcin is a one kind of RIP and stripe et al., first isolated it from the endosperm of *J. curcas* by (Stripe et al., 1976), which have anti-tumour activity and reported some researcher (Lin et al., 2003a; Lin et al., 2003b; Luo et al., 2006). Lin et al., also tested the antitumor activity of curcin by MTT assay. They confirmed cucin inhibited the protein formation in reticulocyte lysate and its mode of action as like as *N*-glycosidase (Lin et al., 2003b). Luo and co-workers also used cucin to inhibit tumour cells growth at dose of 5 μgml^{-1} (Luo et al., 2006).

Zhao and co-workers investigated the effect of curcin on mouse sarcoma-180 cells. They also showed a concentration of curcin (100 μg/ml) inhibited the growth mouse sarcoma-180 cell by over 40% after seven days of incubation (Zhao et al., 2012).

3.9. Antifertility Activity

Crude extract of *J. curcas* have anti-implantation and antifertility effects. In 1995, Goonasekera and co-workers reported the action of fruit extract on pregnancy in the rat. They observed, the crude fruit extracts initiated foetal resorption by interrupting pregnancy incidence at an early stage after implantation. The hot petroleum ether extract degenerated 90% of the foetuses while the hot methanol extract reserved 52% of the foetuses.

Whereas crude dichloromethane extract was inactive at a low dose (Goonasekera et al., 1995). Makonnen et al., also reported antifertility effects of crude seed extract by oral administration to albino pigs (Makonnen et al., 1997). Odusote et al., exposed a notable contraceptive activity of *Jatropha* oil and showed up to 1g/kg doses have promising contraceptive activity without any form of toxicity.

Nonetheless, above 2g/kg doses caused noteworthy acute toxicity in rats and 0.5g/kg doses was chosen as a comparatively safe dose for rats (Odusote et al., 2002). Recently Alwi and Sukardi reported the seeds of *J. curcas* contain 27-40% oil consisting of curcin and tetramethylpyrazine which cause abortion, fetotoxicity and teratogenic effects if consumed during pregnancy (Alwi and Sukardi, 2013).

3.10. Anticoagulant Activity

The crushed fresh leaves and latex of *J. curcas* is used as a haemostatic or a styptic agent in folk medicine to reduce the clotting and bleeding time (Neuwinger, 1996; Watt and Breyer-Brandwyk, 1962). In 1999 Odusote and co-authors reported that the latex of *J. curcas* decreased clotting and bleeding times meaningfully as compared to standard. They observed maximum clotting activity at the latex concentration of 4.7% w/v and 100% v/v (Odusote et al., 1999). Osoniyi and Onajobi also observed latex of *J curcas* decreased the clotting time. However, low concentration or high dilutions of latex protracted the clotting time but didn't clot the blood At the lower concentrations, the ethyl acetate and butanol extract presented a pro-coagulant and anticoagulant activity respectively. Whereas aqueous extract inactive to decrease the clotting time and prothrombin time but marginally prolonged the activated partial thromboplastin time (APTT) (Osoniyi and Onajobi, 2003).

3.11. Hepatoprotective Activity

Balaji and his co-workers assessed the hepatoprotective action of crude methanol leaves extract against hepatocellular carcinoma persuaded by aflatoxin B_1 by oral administration in rats. They observed this extract reduced the levels of elevated serum enzymes, lipid levels and bilirubin and increased the protein and uric acid levels.

It also reduced incidence of liver lesions, lymphocytic infiltrations and hepatic necrosis induced by aflatoxins, which confirmed by the liver histopathology (Balaji et al., 2009b). This methanolic leaves extract of *J. curcas* also tolerated the liver, kidney, and hematological system of rats due to its toxicity (Igbinosa et al., 2013).

3.12. Anthelmintic Activity

Anthelmintics are drugs that are used to treat infections with parasitic worms. In 2006, Jalalpure and co-workers investigated the anthelmintic activity of seed oil against *Pheretima Posthuma (Annelida)*. They observed worm did not show sensitivity on oil sample, but all the worm dies after 42 (Jalalpure et al., 2006). Adamu et al., Reported antischistosomal activity of methanol extracts of *J. curcas* on *S. mansoni* infection in mice and observed this extract reduced only 8.33% of worm (Adamu et al., 2006). Eguale and Giday determined the anthelmintic property of the *Jatropha curcas* seeds against eggs and adult *Haemo nchuscontortus*. They observed seed extracts showed poor activity against live adult parasite, but good activity to hinder the egg hatching (Eguale and Giday, 2009). Ahirrao and co-workers investigated the same activity for aqueous leaf extracts by using adult earthworms *P. posthuma*, while the piperzaine citrate (10mg/ml) and distilled water used as standard and control respectively. They indicated that the leaves of the aqueous extracts (100mg/ml) showed significant anthelmintic activity (Ahirrao et al., 2009). Monteiro et al., evaluated this activity for crude hexane, ethyl acetate (EA) and ethanolic seeds extracts. Tween 80 (5%) and thiabendazole (0.025 g ml^{-1}) used as a negative and positive control respectively. They showed almost complete egg hatching (99.8%) inhibited by the ethanol seed extracts (50 mg/ml) and ovicidal effectiveness reduced to 91.9% after incubation with polyvinyl polypyrrolidone (PVPP), while hexane and ethyl acetate inhibited 15.3% and 32.2% egg hatching respectively (Monteiro et al., 2011).

4. Energy and Food Security

Food is compulsory for every life and energy is for comfortable life, economic growth and industrialisation. Both food and energy security are burning issues for the sustainability of world.

About 805 million people out of 7.3 billion of total world population (one in nine) were hungry (under nourishment) in 2012-2014. One in eight people of developing counties (791 million people out of 805 million) suffer food shortage. Eleven (11) million people are undernourished in developed countries also (FAO, 2014) http://www.fao.org/3/a-i4030e.pdf.

The world energy spending has been increasing, and during the period of 50 years ending in 2025, it will have increased three fold. Over 80% of the total combusted energy is still coming from fossil fuel. The fossil fuel oil is expected to face a greater threat of becoming exhausted. Global oil production may likely reach a maximum between 2015 and 2030.

It is then expected to sharply decline to 40% of today's production by 2075 (Demirbas and Demirbas, 2010).

The world is looking for renewable energy. Solar, wind, tidal, hydrogen, biomass and biofuels are common sources of renewable energy. Biomass and biofuels have been getting attention because of its renewability, sustainability, production easiness and low cost also. Bio-ethanol is a form of biofuels that is produced by fermentation of carbohydrate.

Many food grains and agricultural products that contain sugar, starch or cellulose can be used as raw materials for ethanol fermentation. Brazil and some other developed country have been using food materials for biofuesls. The criticism is the humanity.

Many people in the world are starving at the same time food grains have been draining. Biodiesel another form of biofuels can be produced from vegetable oil. More than 350 oil-bearing crops have been identified, of which only soybean, palm, sunflower, safflower, cottonseed, rapeseed, and peanut oils are considered Vegetable Oils (Goering et al., 1982; Pryor et al., 1983). Table 1 shows the oil species that can be used in biodiesel production.

Uses of edible oils conflict with food security as the use of food grains for bio-ethanol. Non edible oil plants such as jatropha or ratanjyote or seemaikattamankku (*Jatropha curcas*), karanja or honge (*Pongamia pinnata*), nagchampa (*Calophyllum inophyllum*), rubber seed tree (*Hevca brasiliensis*), neem (*Azadirachta indica*), mahua (*Madhuca indica and Madhuca longifolia*), silk cotton tree (*Ceiba pentandra*), jojoba (*Simmondsia chinensis*), babassu tree, *Euphorbia tirucalli*, microalgae, etc. are commonly available in many parts of the world.

Table 1. Oil containing species for biofuel production

Edible oil source	Coconut (Copra), corn (maize), cottonseed, canola (a variety of rapeseed), olive, peanut (groundnut), safflower, sesame, soybean, and sunflower Nut oils Almond, cashew, hazelnut, macadamia, pecan, pistachio, walnut, Amaranth, apricot, argan, artichoke, avocado, babassu, bay laurel, beech nut, ben, Borneo tallow nut, carob pod (algaroba), cohune, coriander seed, false flax, grape seed, hemp, kapok seed, lallemantia, lemon seed, macauba fruit (*Acrocomia sclerocarpa*), meadowfoam seed, mustard, okra seed (hibiscus seed), perilla seed, pequi (*Caryocar brasiliensis* seed), pine nut, poppyseed, prune kernel, quinoa, ramtil (*Guizotia abyssinica* seed or Niger pea), rice bran, tallow, tea (camellia), thistle (*Silybum marianum* seed), and wheat germ
Non Edible oil source	Algae, babassu tree, copaiba, honge, jatropha or ratanjyote, jojoba, karanja or honge, mahua, milk bush, nagchampa, neem, petroleum nut, rubber seed tree, silk cotton tree, tall, castor, radish, and tung

And it is cheap in compare to edible oil sources (Karmee and Chadha, 2005). Jatropha having some degree of advantages among these non edible oil sources can be a good source of biodiesel feed stock. It will maintain both food and energy security of the world population.

5. Seed Oil and Oil Properties

5.1. Oil Extraction

Jatropha seeds contain 40-60% of oil depending on variety (Murali Krishna M et al., 2014; Ong et al., 2011). The first step of oil extraction is the removal of shells from the seeds after collecting the ripe fruits from trees. Seed oil can be extracted manually, mechanically, chemically and ensymetically. Oil extraction process is shown in Figure 2.

Oil can be extracted by mechanical pressure, solvent extraction and enzymatic degradation of kernel. Mechanical extraction yields about 90% of total oil in seed (Subroto et al., 2015). Solvent and enzymatic extraction yields almost 100% of oil in seed.

However, these are complex process and take long time. Solvent extraction involves handling of large volume dangerous chemicals. Commercially Suitable enzyme (s) is still not available for enzymatic extraction of oil from seed kernel (Shivani et al., 2013; Winkler et al., 1997).

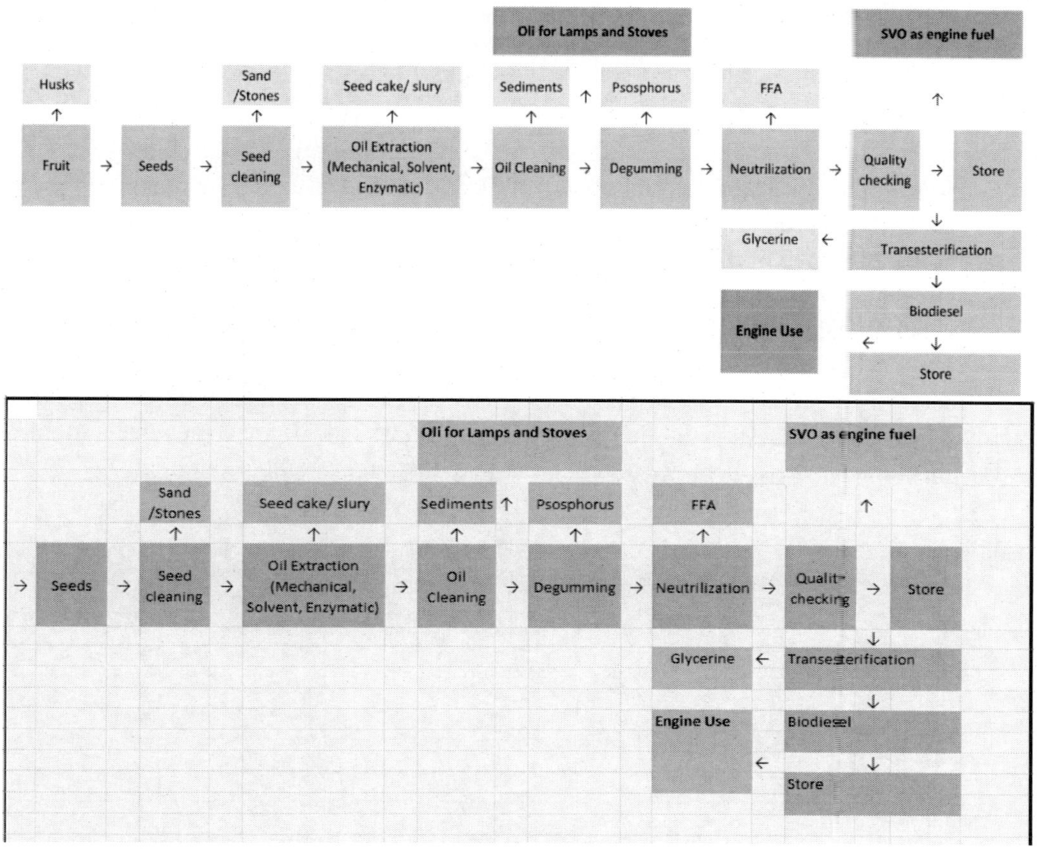

Figure 2. Oil Extraction steps and use (Enclosed with email).

Mechanical process is the use of a machine to exert pressure on seeds for removal of oil. After cleaning and checking the seeds are feed into hopper of machine. For Jatropha seed 0.41 litter of oil is extracted from 1 kg. Mechanical parameters and pretreatments of seeds affect oil yields. The effects of treatment and physical parameters on oil extraction are shown on Figure 3. The amount of oil that can be recovered from the seeds is affected by:

- Throughput: it is the amount of seed crashed per hour (kg/hr). The higher throughput recovers less amount of oil per kg of seeds, because of short time exposure of seeds to pressure. By altering the turning pace of the screw throughput can be regulated.
- Oil point pressure: it is the amount of pressure needed to start oil flow from the seeds. If, it is possible to reduce the oil point pressure, the oil extraction becomes easier.
- Pressure: the more pressure recovers more oil from the seeds. But oil recovery at high pressure experienced more solid particles with oil. It makes the removal of solid particles more difficult. Pressure range of 50-150 bar is the typical operating pressures for engine-driven oil extraction.
- Nozzle size: smaller pore nozzle causes higher pressure and therefore higher oil yield. An optimum is needed for smooth operation.
- Hull content of the seeds: It requires less energy for pressing seeds without hull. And there are no hull fibers in the crude oil. However, it appears like paste inside standard expellers, which sticks to the worm and keeps rotating along with it.

5.2. Seed Oil Properties

Extracted oil quality depends on genetics of plants, environment and soil quality of plants being planted, maturity stage of seeds, extraction methods, presence of contaminant, purification and storage of oil. The physio-chemical properties of Jartopha seed oil is listed in Table 2. In general oil from Jatropha seeds is more viscous. It is relatively low content of free fatty acid but high content of unsaturated oleic and linoleic acids. Jatropha oil also has a high cetane (ignition quality) rating, high content of oxygen and low content of sulphur.

Viscosity is a problem of Jatrpha oil to be used. The low content of free fatty acids improves its storability. High unsaturated oleic and linoleic acids make it remain fluid at lower temperatures where increase susceptibility to oxidation in storage condition.

Pre-treatments		Oil Yield	Pressure	Temperature	Throughput	Energy/litter
heating	▲		▼	▲		▼
boiling		▼	▲	▲		▲
flaking		▼		▼		▼
moisture content	▼	▲	▲	▲	▼	▲
hull fraction	▼	▼	▼	▼	▼	▼
Mechanical factors						
RPM	▼	▲	▲		▼	▲
Restriction size	▼	▲	▲	▲		▲

Figure 3. Pre-treatment and mechanical factors effect on seed oil recovery.

Table 2. Chemical and physical properties of Jatropha oil

Parameter	Jatropha oil
Density at15 °C	0.920gr/cm^3
Viscosity at 30 °C	52cSt
Flash point	240 °C
Fire point	274 ± 3 °C
Cloud point	9 ± 1 °C
Pour point	4 ± 1 °C
Cetane number	38
Caloric value	38.20Mj/kg
Conratson carbon residue	0.8 ± 0.1 (%w/w)
Hydrogen	10.52 (%w/w)
Sulfur	0 (% w/w)
Oxygen	11.06 (%w/w)
Nitrogen	0
Carbon	76.11 (%w/w)
Ash content	0.03 (%w/w)
Neutralization number	0.92 mg KOH/gr
Saponification value	198
Iodine number	94
Monoglycerides	Not detected
Digl icerides	2.7% m/m
Triglycerides	97.3% m/m
Water	0.07% m/m
Phosphorus	290 mg/kg
Calcium	56 mg/kg
Magnecium	103 mg/kg
Iron	2.4 mg/kg

Source: Gubitz et al. 1999.

The presence of high oxygen increases the combustion efficiency. Where, the low sulphur content reduces the emission of toxic gas (SO_2) (Brittaine and Lutaladio, 2010; Gübitz et al., 1999). Oil quality and consistency are important for direct use of oil as fuel and use as biodiesel feedstock. Presence of low contamination in oil, low acid value, high oxidation stability and low contents of phosphorus, ash and water is necessary for good use of Jatropha oil.

6. Energy from Jatropha

6.1. Direct Combustion

Jatropha is a woody plant. Wood, fruit shells, seed husks and kernel are the components of Jatropha from which energy can be derived (Singh et al., 2008). Pruning and cutting down once in every 10 years to a height of 45 cm for allowing re-growth are common practice of

Jatropha cultivation. By pruning every after six years over 20 tons of woody biomass is produced from one (1) hectare of *Jatropha* plantation.

Moreover, cutting over a 10 years period produces 80 ton/ha wood. That can generates 1.2 PJ of energy with an energy content of 15.5 MJ/kg (Sotolongo et al., 2009). The average fruit yield of Jatropha is 3.5 ton/ha (Kumar et al., 2003). Each fruit contains 2-3 seeds of 2.5 cm long on average. Fruits composed of shell, husks, karnels.

About one tone of shell material yields from one hectare after removal seeds from fruits. The shell contains 69% Volatile matter, 15% ash and 16% fixed carbon (R. Singh et al., 2008). The shell caloric value is 11.1MJ/kg (Sotolongo et al., 2009). Considering this value the shell can generate 11.1GJ ha-1. Husks (42%) and kernel (58%) are the constituents of seeds. The seed husks calorific value of is 16 MJ kg-1 that is comparable to that of wood. In gross the seed energy value (24 MJ kg-1) is higher than cow dung and lignite coal (Augustus et al., 2002).

6.2. Seed Oil

6.2.1. Lamps

Lamp uses to enlighten habitat during night is an ancient use by human. Still a considerable number of world population especially Africa and Asia have been using it. Diesel and kerosene are being used as fuel for lamps. Raw Jatropha oil can substitute the conventional lamp fuels. The use of raw oil for lamp has been practiced.

It faced some troublous because of its high viscosity. The lamp become dim as the oil level diminishes so that the oil has to travel longer distances through the wick and the formation of cokes on the wick's surface. Moreover, the ignition temperature of jatropha oil (240 °C) is supposed to three fold higher than petroleum (84°C). It makes the fuel difficult to be ignited. To overcome the problem special design lamp can be used. 'Binga lamp' a special design lamp developed by the binga trees project in Zimbabwe (Figure 4). The lamp use floating wick that keeps the distance between the flame and the oil constant.

6.2.2. Cooking Stoves Oil

People use cooking stoves due to limited availability and accessibility of gas, electricity and firewood. It is also more convenient than the use of firewood. In rural areas, replacement of firewood by plant oil for household cooking will mitigate the problems of deforestation as well as improves the health of rural women who confronted to the indoor smoke pollution from cooking. Jatropha oil performs very satisfactorily when burnt using a conventional (paraffin) wick after some simple design changes in the physical configuration. Three modified stove models are available. UB-16 stove directly uses Jatropha kernels. The The Wheel brand stove, a typical example of an adapted kerosene stove. And 'Protos' utilizes vaporized and sprayed jatropha oil under pressure into a specially designed stove. It is developed by BSH Bosch and Siemens Hausgeräte GmbH (Figure 5).

6.2.3. Direct Use As Engine Fuel

Straight *vegetable oil* (SVO) satisfies the qualities to be used as fuel for compression ignition engines (diesel engines).

Figure 4. Binga lamp for use of Jatropha seed oil.

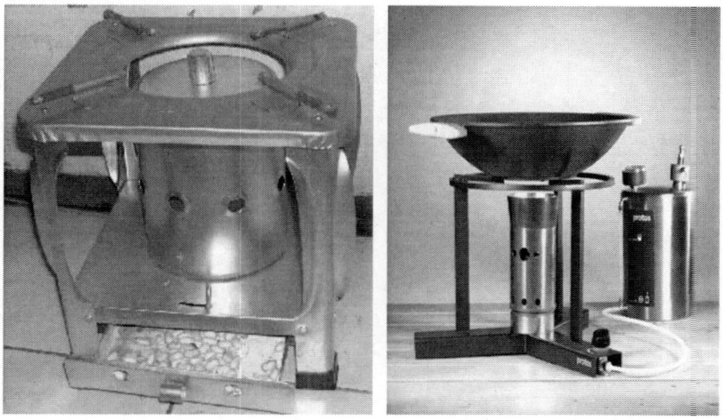

Figure 5. Cooking stoves for use of Jatropha oil and seed Kernels.

Usually, any warm diesel engine will run on heated SVO. Some properties of SVO differ from fossil diesel. These have impact on combustion efficiency, emissions, fuel consumption and life time engines. The problem may be because of deposition of unburned fuel in the engine or having aggressive properties that leads corrosion. Quality of SVO to be used directly as fuel and selection of engine to be run by SVO are important factor to be considered for successful use of SVO.

6.2.3.1. SVO Quality Standard

Different levels of quality require for different uses of Jatropha oil. And the quality question comes when it is thought to be used as engine fuel. In Germany a quality standard of rapeseed oil as engine fuel has been developed (Table 3).

It categorized as characteristics properties that depends on source of oil and variable properties that depends on processing, filtering, treatment etc. the limiting values can also be applied to other oils like Jatropha though this standard have been developed for rapeseed oil.

The value of each criteria of SVO should be in between desirable range. Contaminations (the presence of foreign materials), acid value (content of free fatty acid in the oil), oxidation stability (stability of oil in a hot environment), phosphorus content, ash content and water content are the limiting factors that controls the variable properties of SVO.

SVO content 4%-5% less energy per volume in compare to fossil diesel. This limitation can be overcome by more efficient combustion by molecular oxygen content.

Table 3. Standard SVO quality criteria for diesel engine

Properties / constituents	Units	Standards
Density at 15°C:	900-930 km/m^3	According to DIN EN ISO 3675 / 12185
Flash point: min.	220°C	According to DIN EN ISO 2719
Kinematic viscosity at 40°C: max.	36.0 mm2/s	According to DIN EN ISO 3104
Calorific value: min.	36,000 MJ/kg	According to DIN 51900-1, -2, -3
Carbon: max.	0.40%	According to DIN EN ISO 10370
Iodine value	95-125 g / 100 g	According to DIN EN 14111
Sulphur content	10 mg/kg	According to DIN EN ISO 20846 / 20884
Total contamination	24 mg/kg	According to DIN EN 12662
Acid number	2.0 mg KOH / g	According to DIN EN 14104
Oxidation stability at 110°C: min.	6.0 h	According to DIN EN 14112
Phosphorus content: max.	12 mg/kg	According to DIN EN 14107
Total magnesium and calcium: max.	20 mg/kg	According to DIN EN 14538
Ash content (Oxidasche): max	0.01%	According to DIN EN ISO 6245
Water: max.	0.08%	According to DIN EN ISO 12937

Naturally SVO contain 11%-12% of oxygen in its molecular structure. High viscosity and flash point of SVO make challenging to start cold engine and efficient combustion until the engine become hot.

6.2.3.2. Engine Selection and Conversion

Worm diesel engine can run well using heated SVO. The main obstacles of using SVO are to get the engine started and clean combustion until normal operation temperature (80 – 90°C) is reached. For safety running using SVO, the typical diesel engine should be converted to adapt the different properties of SVO. Introduction of a heating system to reduce the viscosity of SVO and modification of injectors and glow plugs to be start the engine with SVO are main targets.

There are many different types and sizes of diesel engine. Most of them can be modified to SVO running system. The most important factors to be considered for conversion are injection pump and cooling system of the engine. Other factors are load pattern, preheating system, lift pump. Evaluation of physical condition of engine is also important. Engine should be well adjusted and well-maintained condition. If the engine smokes more and the injectors are worn or the glow plugs burned out, before conversion these problem must need to fix. The engine should have a well thermostat system to maintain the optimum operation temperature. If there is no or defected thermostat, the engine might be cool. Therefore, suitable temperature may not reach specially at low load.

According to fuel injection system engines can be categorized by direct injection (DI) and indirect injection (IDI). For DI engines fuel is directly injected into relatively large and cold cylinder. Fuel is injected into small and hot pre-chamber in case of IDI engines. Form here fuel combustion starts before the final combustion into the cylinder. IDI engines are more suitable for conversion to SVO. DI engines are more sophisticated to fuel quality and load pattern. Thus, it requires more attention and special fuel delivery (2-tank) system to be converted to SVO.

The main problem with DI engine is the deposition of unburned SVO on the piston and piston rings, finally reached it to crankcase through cylinder wall. When it reached to crankcase it become diluted with lube oil. The concentration of SVO in lube oil always increases because it does not evaporate due to its high boiling point. If the concentration reaches > 10% SVO in lube oil there will be polymerization caused by thermal load of the mixture. The polymerization causes sudden and dramatic increase the viscosity of lube oil and the engine may damage or totally destructed (Thuneke et al., 2005).

For efficient use of SVO and engine longevity, it is better to modify the engine to withstand SVO by skilled technician. There are two main ways of modification.

- Single tank system: this system enables IDI engines to be started directly on SVO. The important modification of this system is the presence of a glow plug in the combustion chamber. Installation of glow plugs and injectors are the requirements. Adjustment of injection pressure and timing are curtail. It is the same of using single tank SVO and original diesel engine. Cold start is the only difference. Single tank SVO engine cold start requires allowing the pre-heating work 5-10 second more than typical diesel engine and the adjustment of gas with accelerator. It is often recommended for IDI engines.
- Double tank system: generally DI engines are modified to double tank system. An extra fuel filter, fuel tank, fuel heating system, two ball valves some horses are the requirement of conversion of a basic diesel engine without electric starting, preheating and electrical control of fuel switching. The cortical points to design this system are shortening the purging time and to ensure not to mix the SVO with diesel in diesel tank during purging process. This system engine started on diesel and switched to heat SVO manually or automatically when the engine temperature have rise. For next time smooth starting, it is important to switch to diesel in advance before stopping the engine for cooling. Dual tank modification is not suitable for that engine having many starts/stops, idling/low load or only running for short distances. It is not convenient to use because of frequent switching and regular checking the fuel of both tanks. By delaying the return value, mixing of SVO to diesel in diesel tank can be prevented. But diesel consumption increases by this modification. Another way of mixing prevention is to loop the return fuel to injection pump rather than the diesel tank. It increases purging time.

Another way of using SVO as direct fuel is blending with diesel. There is no need of modification of typical diesel engines. Diesel reduces the viscosity and flash point as it mixes easily with SVO. IDI engines can run properly up to 50% blending. Blending fuel is not suitable for DI engines. Maximum 20-30% can run the engines. Initially it seems the engine works well but in long run it leads serious irreversible damage of the engine. It causes by the deposition of unburned SVO.

6.2.5. Biodiesel Feedstock

Jatropha seed oil can be directly used by modifying the engine. Biodiesel, the chemical modification of seed oil can also be used by engine directly. One of the main targets of Jatropha utilization is biodiesel production from seed oil.

Petro-diesel properties are very close to biodiesel properties (Table 4). Biodiesel gives better ignition and combustion. It also minimizes the emition of harmful smoke, sulfur and nitrogen oxide (Palash et al., 2014). However, biodiesel contain less energy that may causes 2-10% more fuel consumption. Biodiesel acts as solvent and for the first time when it has been putted into fuel tank of a diesel engine, it may causes filter blocking because of the cleaning of the dirt of pervious diesel (Sorate and Bhale, 2015). Special gaskets are needed to use to retain the integrity of fuel line of biofuel engine.

Transtarification is the chemical term for the production of biodiesel from SVO. In this process, biodiesel (ester) forms by the reaction between fat/oil (triglyceride) and alcohol in presence of catalyst (Figure 6). Here glycerine, water with soap, recaptures alcohol and free fatty acids are produced as byproducts. These byproducts can be reused. The quality of the biodiesel is influence by the composition of seed oil, chemical and physical impurities. climate and soil of the area of cultivated seed may affect the Physio-chemical properties of oil (Islam et al., 2012). Particles and sediments are the physical impurities that can be removed by filtration. Chemical contaminant is not a problem. If the oil is being stored for a long time under unfavorable condition, it needs to check acidity (FFA) and water content before reaction.

Any alcohol can be used. Generally, methanol is used due to its advantage over others. Three types of catalyst (base, acid and two-step acid) can be used for transterification reaction. Sodium hydroxide (NaOH) or potassium hydroxide (KOH) is often used as base catalyst, because of having some degree of advantages over other catalysts.

Table 4. Comparative physio-chemical characteristics of standard diesel, seed oil, and converted biodiesel of Jatropha

Physio-chemical characteristics	Standard diesel	Jatropha oil	Jatropha biodiesel	Limits
Specific gravity	0.82–0.89	0.90–0.92	0.85–0.90	0.86–0.90
Viscosity (40 °C)	3.7–5.8	24.5	5.2	1.9–6.0
Cetane number	46–70	23–41	57–62	47 min
Sulphur (%)	< 1.0–1.2	0.13–0.16	< 0.02	20 max
Density at 15 °C (g/cm^3)	0.84–0.89	0.92	0.87	0.87–0.89
Cloud point (°C)	5	11	8.2	–
Pour point (°C)	−20	4	2	–
Iodine number	60–135	90.8–112.5	95 to 106	100–120
Neutralization number (mg KOH/g)	≤ 0.8	0.92	0.24	–
Flash point (°C)	68	225	135	130 min
Calorific value (MJ/L)	42.25	39.66	42.67	35 min
Peroxide value (meq/kg)	–	4.6	–	–
Saponification value (mg KOH/g)	–	196	–	–
Refractive index	1.32	1.46–1.61	0.92–1.46	–
Acid number (mg KOH/g)	0.104	1.0–38.2	0.4	0.80 max
Solidifying point (°C)	−14	2	−10	–
Boiling point (°C)	248	286	255	–

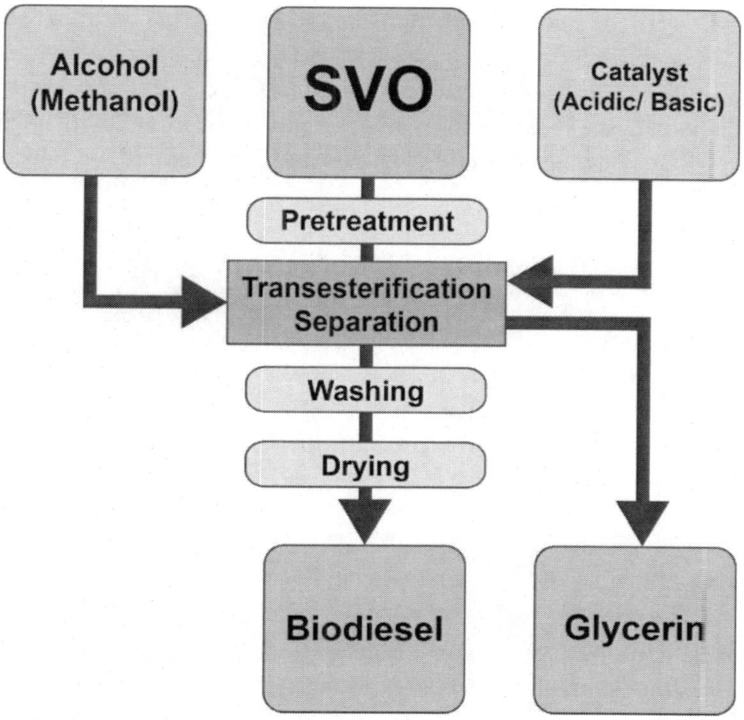

Figure 6. Chemical conversion of vegetable oil (SVO) to Biodiessel – Transesterification.

Base catalyst has high conversion rate, rare side reaction and less reaction time. It can work low temperature (150 F) and pressure (20 psi) no use of exotic materials and direct conversion to methyl ester. El Diwani et al. (2009) obtained 97.6% (wt) biodiesel after transesterification of 100 kg seed oil using methanol and NaOH as catalyst.

6.3. Biogas

The renewable energy biogass is the mixture of gasses that are produced by the fermentation of biological wastages. The main component of biogass that produces energy is methane (CH_4). This gas can be used for cooking, lighting, electricity generation etc. Seed cake, fruit shell and vegetative plant of Jatropha can be an excellent source of raw material along with other biological residues. Foidl and Eder (1997) used seed cake as a single digestion input for biogas production (Foidl and Eder, 1997). Biogas production efficiency of jatropha residues is summarized in the (Table 5). Moreover, the slurry from biogass plant is a good source of organic fertilizer (Yadav et al., 2014).

6.4. Biohydrogen

Hydrogen (H_2) gas became very attractive renewable energy for future because of its less energy expenditures and the possibility to use low cost organic wastes as substrate. Jatropha Seed cake can be used as substrate hydrogen (H_2) production.

Lopes et al. (2015) studied dark fermentation by a pure strain of the bacteria *Enterobacter aerogenes* (Lopes et al., 2015). They produced 68.2 mL H_2/gVS_{iJSC} biohydrogen without pretreatment of substrate. It is significant in the viewpoint of energy saving. They found that, the increase concentration press cake from 2.5 to 10 gVS/L_{FM} led to the increase of the cumulative hydrogen production and to higher $bioH_2$ production.

7. JATROPHA FOR AGRICULTURE

The small plant Jatropha bears importance in agriculture. The seed cake, fruit shell, leaves and other residues of plant that are very good as organic manure. The seed cake can also be used as animal feed as it contains high protein. Oil and extracts have some pesticide and herbicide properties that are useful to control pests and herbs. Use of jatropha plant as hedge is an ancient application.

7.1. Organic Manure

The seed cake and other residues of Jatropha contain Nitrogen, Potassium, Phosphorus, Calcium, Magnesium, Zinc, Iron, Cupper, Manganese, Borne, and Sulfur (Kumar and Sharma, 2008). These macro and micro nutrient can be an excellent source of soil nutrient and soil conditioner. The value of macronutrient present in seed cake is presented on Table 6. NPK content of Jatropha is similar to neem oilcake, castor oilcake, cow dung, and poultry litter (Islam et al., 2012). There is significant increase of millet yield in Mali by using Jatropha seed cake as fertilizer (Henning et al., 1995). Moreover, leaves of increase the microbial activity (Gübitz et al., 1999).

7.2. Pesticides

The extracts from jatropha root, latex, bark, leaves, fruit, seed, and oil show antimicrobial activity against pest, rodent, fungus, bacteria, Mullica (Consoli et al., 1989; Grainge and Ahmed, 1988; Jain and Trivedi, 1997; Meshram et al., 1994). Undesirable effects of extracts on human, cows, rodent, sheep, and poultry are also reported (Agaceta et al., 1981; Devappa et al., 2010; Goel et al., 2007). Jatropha extract is effective against bollworm, a harmful insect of cotton, pulses, potato, maize (Kaushik and Kumar, 2005).

Table 5. Production of biogas from jatropha residues

Jatropha byproduct	Inputs (Kg)	Biogas yield
Seed cake	800	360 m^3
Oil residue	800	360 m^3
Fruit shell (fresh)	3000	100 m^3

Table 6. Macronutrients of Jatropha seed cake

N%	P%	K%	Ca%	Mg%	Source
4.4-6.5	2.1-3.0	0.9-1.7	0.6-0.7	1.3-1.4	Achten et al. (2008)
3.0-4.5	0.65-1.2	0.8-1.2			Patolia et al. (2007)
4.91	0.9	1.75	0.31	0.68	Wani et al. (2006)

To protect seedling, fruit bunches of oil palm from rat, rhinoceros beetles and insects; a blend of blend of neem oil, jatropha oil, and turpentine oil (extracted from pine trees) has been using in Malaysia (Srivastava et al., 2008).

7.3. Herbicides

Weeding is a problem in agriculture. Generally weeds are uprooted by physical pulling using garden tools, metals stones etc. it is laborious and time consuming. Some chemicals like glycophosphates show herbicidal activity. These chemicals have been using for weeding of crops. Jones and Csurhes (2008) found that fatty acids and esters of fatty acids are the component of herbicidal activity (Jones and Csurhes, 2008). As Jatriopha oil is rich of fatty acid it could be use as active ingredient of herbicide preparation.

7.4. Bio-Fence

Farmers use different materials to protect their crops from animals. Use of Jatropha plant as hedge is ancient. The advantage of using Jatropha tree as bio-fence is somehow permanent and animal doesn't feed on it (Henning, 2009). It is small tree and can be managed by cutting. Jatropha plantation at 5 cm distance can prevent penetration even chickens. The same reason it can also be used as homestead boundaries.

7.5. Sericulture

Sericulture is the production of silk by culturing silkworm. The species *Bombyx mori* is commonly used. Silkworm larvae feed on mulberry leaves and produce the cocoons. Jatropha leaves can be an alternative feed for larvae as it can feed on it and able to produce cocoons (Reyadh, 2004).

7.6. Animal Feeds

Jatropha seed cake bears high (50–62%) protein content and trace amount of amino acids (Makkar et al., 1998). This can be an excellent source of crude protein in animal feed.

The problem of using Jatropha seed cake as feed is the presence of toxic component that can causes health hazard of animals.

Detoxification of seed cake is possible by treatment (Guedes et al., 2014; Nazir et al., 2014). After detoxification it can be used as feed supplement (King et al., 2009; Makkar et al., 2008). There are nontoxic varieties of Jatropha in Mexico and Central America. Seed cake from these plants can be directly used as animal feed.

8. Environment Protection

Sustainable environment for future generation is a burning issue now-a-days. Economic growth, global warming, emiton of toxic gases, deforestation are always heating to world leaders. Cultivation and use of Jatropha is environment friendly. Biodiesel from Jatopha can minimize the dependency on fossil fuel as well as global warming. Biopolymers from Jatropha are biodegradable. Jatropha plant itself can be used to control of soil erosion and for bioremediation of biological wastes.

8.1. Soil Erosion

Soil erosion is a problem in unplain land and costal area. Hedge of Jatroha is effective to control soil erosion. Jatropha produces taproot that fix the plant tightly with soil. The profusion of lateral and adventitious roots near the surface binds the soil and keeps it from being washed out by heavy rains (Henning, 2009). By reducing wind velocity and binding the soil with their surface roots jatropha hedges reduce erosion by wind. These anti erosion capacity depend on rain fall and wind velocity. If there is heavy rain and high velocity wind it is less effective.

8.2. Protection from Wind

Gusty wind demolishes the habitats, trees and other establishments especially in coastal areas. Jatropha plant fix with soil very well and it can grow on a diverse land. A biological barer can be establish by plantation of Jatropha at low distance at coastal areas. It can reduce the velocity of wind.

8.3. Bioremediation of Wastage

Industries i.e. tanneries, papermaking, refining ores, etc. discharge effluents that contain toxic components and heavy metal ions (Ni, Cr, Hg, Cr, Se, Zn, Pb, etc.). All of these are harmful for environment, even for human. Bio-magnification of these metal and toxin is a major concern. The ability of Jatropha to grow on diverse and adverse condition makes it a bioremediation agent. Jatropha plant grown on industrial effluent can detoxify the toxin, uptake and accumulate the metal ions into their biomass. Thus the effluents become environmental friendly. Moreover, seed cake can absorb heavy metal from environment (Abidin et al., 2014).

Garg et al. (2007) studied the adsprotion of cromium (Cr) and cadmium (Cd) from aqueous solutions under various experimental conditions using various agricultural by-products including jatropha oil cake. They found high adsorption by Jatropha seedcake in compared to corn cob and bagasse.

8.4. Bio-Polymers

Every day many products from polymer are being used. The polymerization reaction uses many hazard chemicals. These synthetic polymers are not environmental friendly. To overcome these problems biopolymers draw more attention because of the use of renewable natural components, availability and biodegradable property (Harjono and Sugita, 2012). Polymers such as polyurethanes, polyesters, polycarbonates, etc can be made by using the natural oil polyols. It is possible to use Jatropha oil as raw material for the production of natural polyol (Timothy et al. 2007). Reacting between unsaturated or hydroxyl-group-containing fatty acids or their esters and bi-functional ester or amide-forming compounds produces thermoplastic material. Jatropha seed oil FFA can be one of the reaction materials to produce plastics.

8.5. Carbon Sequestration

Photosynthesis is the biochemistry by which plant uptake CO_2 and produce carbohydrate as energy source. Naturally, CO_2 is balanced in environment by this assimilation process. Environment pollution and global warming is caused mainly due to excess CO_2 emition simultaneously deforestation. Carbon credit gain is possible through Jatropha cultivation (Figure 7). Reforestation is one of the credit origin by Jatropha is planted on less agriculture important lands. Jatropha can grow well diverse land and retain its biological activity.

Each Jatropha tree should sequester a minimum of approximately 0.5 tones of CO_2 in its 40 year life span. Biodiesel, biomass, biogas are other credit origins. These can replace the traditional fossil fuel to renewable natural sources. The organic matter from Jatropha increase soil carbon also. The 20-year chronosequence of *J. curcas* living fences exhibited a significant SOC increase of 62 ± 23 g m^{-2} y^{-1} (Baumert et al., 2014).

9. Industrial Uses

9.1. Soap and Cosmetics

The viscous oil from Jatropha fruit can be used for soap making (Openshaw, 2000). The high level of palmitic acid and the hydrophobicity of Jatropha oil make easy of manufacturing soft and durable soap. It is being used in West Africa, Zambia, Tanzania and Zimbabwe as a soapstock, including in the manufacture of soft laundry soap (Pratt et al., 2002). Jatropha oil gives a very good foaming, white soap with positive effects on the skin partly due to the glycerine content of the soap (Henning, 2000).

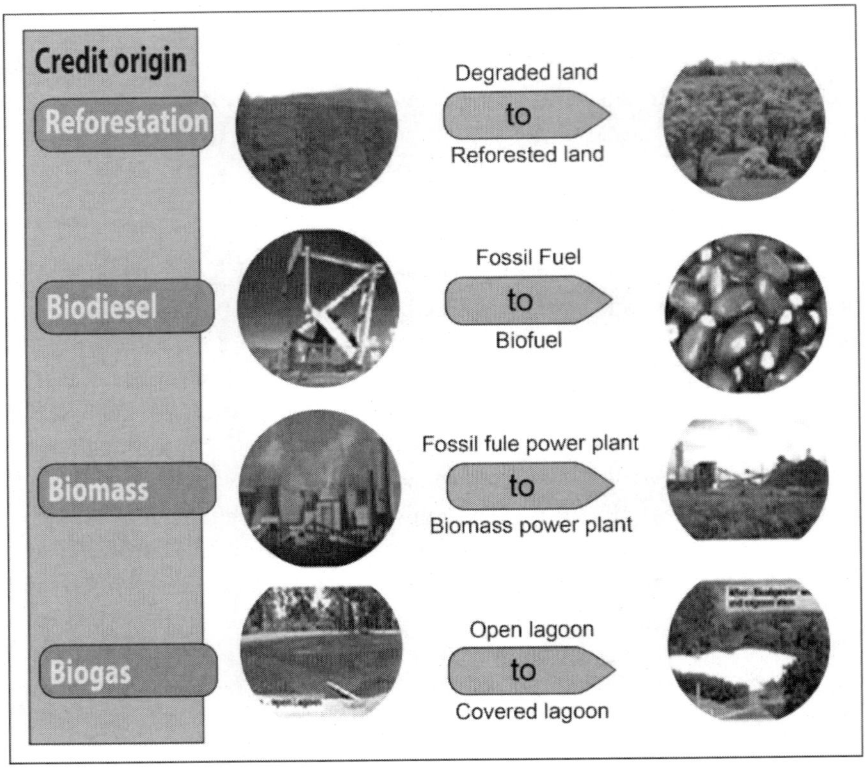

Figure 7. Jatropha plantation and possibilities of gaining Carbon Credit.

Jatropha soap is said to have medicinal characteristics and is therefore used by people with various skin diseases and sensitivity to regular soap (Messemaker, 2008). The 36% linoleic acid (C18:2) content in Jatropha kernel oil is of possible interest for skin care (Benge, 2006; Pratt et al., 2002). The oil is also an ingredient in hair conditioners (Brittaine and Lutaladio, 2010).

9.2. Medicine

Many phyto-chemical constituents, alkaloids, coumarins, flavonoids, lignoids, phenols, saponins, steroids, tannins, and terpenoids, were detected in different extracts from different parts of this plant (Zhang et al., 2009).

Pharmacological studies have demonstrated significant action of different extracts and/or isolated compounds as antimicrobial (Ravindranath et al., 2003), anti-inflammatory (Apu, Hossain, et al., 2012; Bhagat, Ambavade, Misar, and Kulkarni, 2011; Reena, 2011), healing, homeostatic (Oduola et al., 2005), anticholinesterase (Feitosa et al., 2011; Singh and Singh, 2005), antidiarrheal (Apu et al., 2012; Félix-Silva et al., 2014; Silva et al., 2011), antihypertensive (Abreu et al., 2003), and anticancer agents (Kharat et al., 2011; Shahwar et al., 2010). Toxicological studies associated with phytochemical analysis are important to understand the eventual toxic effects that could reduce its medicinal value.

9.3. Dye

The extract from Jatropha leaves, bark of tender steam yields dark blue dye which is used for colouring cloths, finishing nets and lines in the Philippines (Gübitz et al. 1999). The dye may be extracted and concentrated to yellowish syrup or dried to blackish lumpy mass. The dye imparts to cotton different sheds of tan and brown which are fairly fast.

9.4. Thermo Stabilizer

Jatropha seed oil act as thermal stabilizer on polyvinylchloride (PVC). Okieimen and Sogbaike (1996), investigated thermal degradation of PVC at 170, 180, and 190 °C under oxidative and non-oxidative conditions in the presence of Jatropha seed oil. The result viscosity measurements and level of unsaturation revealed that the oil soaps of Jatropha have good stabilizing effect on thermal degradation of PVC under oxidative and non-oxidative conditions.

9.5. Tanning Agent

The chemical Chromium (III) sulfate [$Cr_2(SO_4)_3.12H_2O$] is used as tanning agent in tannery. But the chemical is environmental hazard. The heavy metal persists longer in effluent. From the pre-historic period the tanning of bark, wood, roots, or berries has been using traditionally (Okuda, 1995).

Tanning acts like dehydration of the middle space of the protein fibers and cement of the fiber of hides. It is an active component of Jatropha bark extract. *Jatropha curcas* seed oil can be used as co-stabilizing agent for skins and hides in tanning process. These can be used for leather processing, textile, and coating industries (Sundar et al., 2013).

9.6. Activated Carbon

Activated carbon acts as adsorbent for purification and filtration. It was first introduced in industry at early of 20th century for sugar refining. Since then, the applications of activated carbon have extended to textile, cosmetics and water purification industries. Activated carbons are prepared by physical and chemical activation methods. Almond shells, apricot and peach stones, maize cob, eucalyptus bark, linseed cake, linseed straw, saw dust, rice hulls, olive stones, cashew nut hull, cashew nut sheath, coconut shells and husks, tea waste ash are the agricultural by-products and waste materials used for the production of activated carbons. Besides these, sulfonated coal, tyre coal dust, activated bauxite, cement kiln dust, shale oil ash, ground sunflower stalk, etc are other sources of activated carbon. All these activated carbons have been successfully used for the adsorption processes (Karthikeyan et al., 2008; Ramakrishnan and Namasivayam, 2009; Yang et al., 2010). Seed shells of Jatropha are removed at early stage of oil extraction. It bears toxic and demands detoxification for human and environmental safety.

Table 7. Jatropha seed shell waste activated carbon properties

Properties	HCl	H_2SO_4	$ZnCl_2$	Na_2SO_4	Na_2CO_3	$CaCO_3$	$CaCl_3$	H_2SO_4+ $NH_4S_2O_8$	H_2SO_4+ H_2O_2
pH	9.55	4.67	6.6	8.5	5.74	9.2	7.16	6.85	7.05
Moisture content, %	11	2	8	10	10	16.4	8.2	9.6	10.1
Ash content, %	7.42	11.94	19.57	17.77	14	10.57	13.07	8.67	10.85
Volatile matter, %	30	26.7	27.5	31.2	20	24.2	27.8	31.3	34.2
Fixed carbon, %	69	73.5	59.8	73.1	73.8	72.1	69.7	81	84.3
Conductivity, mS/cm	0.22	0.1	0.61	0.13	0.33	0.37	0.24	0.19	0.42
Specific gravity, S	1.25	1.33	1.49	0.89	1.1	1.48	1.32	1.88	1.37
Bulk density, D	0.44	0.39	0.44	0.4	0.63	0.29	0.39	0.28	0.45
Porosity (S-D/S) 100, %	44.2	53.08	69.48	62.56	64.55	64.86	72.15	61.26	72.34
Matter soluble in water, %	1.84	1.06	1.74	2.12	1.41	1.22	0.72	1.02	1.2
Matter soluble in acid, %	1.25	1.08	1.62	1.42	0.94	0.74	0.92	1.41	1.91
Surface area, m2/g	1195	709	590	456	629	408	1064	664	751
Iodine number, mg/g	1143	668	899	412	589	368	998	612	723
Sodium, w/w %	15	5.5	8	15	5.5	8	6.4	5.5	8.4
Potassium, w/w %	10.6	5.7	3.1	6.1	2	5	5	7.6	7.1
Iron content, w/w %	1.4	1.2	1	0.9	1.1	1.2	0.8	1	1.2
CCl4 activity	0.36	0.55	0.59	1.14	1.19	0.52	0.72	1.32	1.19
Phenol adsoption capacity, mg/g	5.21	2.85	3.62	1.22	2.28	0.8	4.2	3.08	2.64
Yield, %	42	39	44	32	55	37	52	62	32

The best uses of seed shells can be, use as a possible precursor for preparation of activated carbon. Karthikeyan et al. (2008) have chemically activated the carbon from Jatropha seed shell (Karthikeyan et al., 2008). The result (Table 7) disclosed that probably due to more branching and porous carbon with more void space there is a low bulk density value. All the carbons except H_2SO_4, $ZnCl_2$, Na_2CO_3 and $NH_4S_2O_8$ are found to be basic in nature. The quantity of volatile matter shows a higher trend because of the presence of highly porous organic matter. However, High value of ash and volatile matter reduces the quantity of fixed carbon (Karthikeyan et al., 2008; Ramakrishnan and Namasivayam, 2009). Reactive dye, RBBR from textile effluent was removed using activated carbon from *J. curcas* pods by Palanivel et al. (Sathishkumar et al., 2012).

9.7. Bio-Lubricants

Mineral oils, synthetic oils, re-refined oils, and vegetable oils are the main sources of lubricant. Lubricants derived from petroleum oil are mostly available in the market. Which are not adaptable with the environment because of its toxicity and non-biodegradability (Adhvaryu et al., 2005; Salih et al., 2012).

Vegetable oil can be minimize or substitute the use of petroleum lubricant. Renewability, environmentally friendly, biodegradability, less toxicity are the plus point of vegetable oil over petro-lubricant (Kalam et al., 2012; Siniawski et al., 2007). The main limitations of vegetable oil are its poor activity at low temperature, oxidation and thermal stability and gumming effect (Mofijur et al., 2012; Nagendramma and Kaul, 2012). The non edible Jatropha seed oil can be used as bio –lubricant. Blending of 10% jatropha oil with base lubricant can be used as lubricating oil without any shortcoming, which would help to reduce the global demand of petroleum based lubricant substantially (Shahabuddin et al., 2013).

9.8. Adhesives

Fossil based adhesives are generally used. Proteins based adhesives can be a sustainable alternation of fossil based adhesives (Nordqvist et al., 2012). Adhesives, coatings, and surfactants are the most common technical application of industrial proteins (Ramesan, 2005; Vaz et al., 2003). Jatropha leaves contain protein and seed cake is high content of protein that can be a potential source of raw material of industrial protein (Lestari, Mulder, and Sanders, 2011). Properties of Jatropha seed cake protein as adhesive are given on Table 8. Methods of protein isolation from press cake is established (Hamarneh et al., 2010; Lestari et al., 2011; Roach et al., 2012; Vaz et al., 2003). It needs attention to develop it for industrial application.

9.9. Coatings

The most important Alkyd resins contribute about 70% of total used conventional surface coating today. It is low cost because of using relatively less expensive raw material. Moreover, it is highly soluble inexpensive in solvents. Generally alkyd resins are prepared by condensation polymerization of fatty acids or natural oils with polyhydric alcohols such as pentaerythritol, glycerol or trimethylol propane and dibasic acids such as phthalic anhydride and isophthalic acid (Kumar et al., 2010). Decorative paints, industrial paints, weather work paints, air drying and stoving paints, inks, machine toll finishes, and marine topside are common application of Alkyd resins. Patel et al. (Patel et al., 2008) investigated the Alkyd Resin made by polymerization of Jatropha and Rape seed oils with glycerol, phthalic ,and maleic anhydride. They characterized the resin according to physical and chemical properties also compare with commercial resin (syntholacl-60). The result revealed that resin from Jatropha oil is comparable with commercial one and can be used it as natural source.

10. Contribution of Jatropha and Expectation from Jatropha

Having enormous potentials on social, agricultural, environment, sustainable energy production and industrial potentials, Jatropha became attractive to researchers and policy makers. Research showed that Jatropha seed contain 40-60% oil content And productivity 0.1 to 12 tonnes per ha (Heller, 1996; Openshaw, 2000).

Table 8. Jatropha seed protein adhesive properties

Adhesive formulation	13% Casein	13% DOPC protein	13% PC protein
Solvent	Water	0.055M NaOH	0.55M NaOH
Viscosity [cP]direct	888 ±78	931 ± 115	4338 ± 442
Viscosity [cP]2 days	116 ± 14	383 ± 6.2	170 ± 3.5
pH	7	8	8
Color	Yellow	Brown	Brown
Set time (min)	2.5 ± 0.7	1.25 ± 0.4	2.25 ± 0.35
Open time (min)	3.5 ± 0.7	8 ± 2.83	9.5 ± 0.7
Adhesion (60%RH)	+++	+++	++
Adhesion (water)	-	-	+

Source: Lestari et al. (2011).

Yields both fruits and oil are dependable to species, accession, soli, climate and agronomy (Behera et al., 2010; Gopinathan and Sudhakaran, 2011). Seed yield of 4–5 mg/ha/yr is expected to be commercially viable for a Jatropha-based biofuel program.

Containing 30–35% oil and an average seed yield of 3.75 mg ha/yr is economically more beneficial to the average yield profile of soybeans and rape- seed (Singh et al., 2014).

However, the actual seed production of Jatropha in field condition was poor than expectation (Bailis and Kavlak, 2013; Everson et al., 2013; Singh et al., 2013; Tikkoo et al., 2013). The reported yields of Jatropha in field condition in India, Belgium, South Africa. Tanzania, are 0.5–1.4 mg/ha/yr, 0.5 mg/ha/yr, 0.35 mg/ha/yr, 2 mg/ha/yr respectively (Kant and Wu, 2011). The less productivity is because of unavailability of suitable high yielding varieties, large scale plantation without evaluating the genetic potential of planted materials, consideration of Jatropha as no/low impute crop, lack of knowledge on agronomy.

Jatropha seed cake is an excellent source of protein. To add commercial value it is expected to utilize the press cake as animal feed protein supplement. Presence of toxic compounds is the barer to be utilized for this purpose. Good thing is the discovery of nontoxic Jatropha varieties and the detoxification process of toxins.

Many biological active chemical compounds are extracted from bark, leaves, roots that are expected to be used in pharmaceutical industry. However, toxicity must be studied before the use of Jatroha products as therapeutic agents or medicines. Oil from Jatropha has ancient uses as cooking fuel, soap and cosmetic manufacturing. It is reported that Jatropha oil can be utilized as fuel with diesel engine directly or slight modification of the engine. Jatropha oil can be converted to biodiesel by chemical reaction called transesterification. This biodiesel is favorable to use directly to engine or to blend with petro diesel. High viscosity is the major problem of using Jatropha seed oil as fuel.

11. Technological Intervention for Jatropha Improvement

Jatropha bears multi-dimensional potentials (Islam et al., 2014). But it is still behind to compete to be commercialization. High yielding Jatropha varieties is not found yet.

Lack of agronomic knowledge, unawareness or knowledge gap of farmers and common belief as no impute crop render Jatropha's productivity (Divakara et al., 2010; Edrisi et al., 2015). It requires extensive searching of natural germplasms and systematic breeding programs for genetic improvement. The success of breeding depends on availability of diverse germplasms (Heller, 1996). However, some study showed narrow genetic diversity among the world wide population (Camellia et al., 2014; Ke et al., 2011; Rosado et al., 2010; Satyawan and Tasma, 2011; Sun et al., 2008; Tanya et al., 2011). It limits the success of conventional breeding. Inter-species and inter-genic hybridization, haploid breeding (anther, pollen, ovary culture), somaclonal and germaclonal variation, mutation breeding are biological techniques that have proven records for variation creation and breeding of many important crops. For jatropha there are some reports on organogenesis only. Other technologies are still unexplored. Genetic resources technology i.e. marker assisted selection (MAS), molecular breeding, genomic selection (GS), genome wide association studies (GWAS) and genetic engineering have been using with confident for many crop breeding programs. Jatropha genomic technology research is still far behind in compare to other important crops though some reports are available on that (Jaganath et al., 2014; Kim et al., 2014; Maravi et al., 2015; Patade et al., 2014; Tao et al., 2014; Ye et al., 2014; Ye et al., 2014). Now it is time to explore full potentials of Jatropha by using technologies.

CONCLUSION

The plant Jatropha bears enormous potentials. Jatropha don't compete with food security and it contains high percent of oil also it can grow relatively less impute. For this reasons Jatropha has been getting much attention to researchers and policy makers. However, the previous experience did not satisfy the expectation because of poor productivity of planted Jatropha. It does not decline the potentials of Jatropha. Moreover, it concentrates focusing on high yielding variety development, standardize agronomic practice for maximum yields. For value addition it is also necessity to develop non toxic commercial variety or economic detoxification of toxins in seed cake and utilization of all products and byproducts. Biotechnology, Biochemistry, Biochemical engineering and mechanical engineering can ensure the efficient utilization of Jatropha.

REFERENCES

Abd-Alla, H. I., Moharram, F. A., Gaara, A. H., El-Safty, M. M. 2009. Phytoconstituents of *Jatropha curcas* L. leaves and their immunomodulatory activity on humoral and cell-mediated immune response in chicks. *Zeitschrift für Naturforschung. C, A Journal of Biosciences*. 64(7): 495.

Abidin, Z. Z., Salleh, M., Harun, M., Bakar, A. 2014. Biosorption of Zn (II) from aqueous solution by jatropha curcas press cake. *Journal of Scientific and Industrial Research*. 73 (3): 191-194.

Abreu, I. C., Marinho, A. S., Paes, A., Freire, S. M., Olea, R. S., Borges, M. O., Borges, A. C. 2003. Hypotensive and vasorelaxant effects of ethanolic extract from *Jatropha gossypiifolia* L. in rats. *Fitoterapia*, 74(7): 650-657.

Achten, W., Verchot, L., Franken, Y. J., Mathijs, E., Singh, V. P., Aerts, R., Muys, B. 2008. Jatropha bio-diesel production and use. *Biomass and Bioenergy*. 32(12): 1063-1084.

Adamu, S., Kela, S., Suleiman, M. 2006. Antischistosomal properties of extracts of *Jatropha curcas* (L) on Schistosoma mansoni infection in mice. *Afr. J. Tradit. Complementary Altern. Med.* 3: 37-41.

Adhvaryu, A., Liu, Z., Erhan, S. 2005. Synthesis of novel alkoxylated triacylglycerols and their lubricant base oil properties. *Industrial Crops and Products*. 21(1): 113-119.

Agaceta, L., Dumag, P. and Batolos, J. 1981. Studies on the control of snail vectors of fascioliasis; molluscicidal activity of some indigenous plants. *NSDB Technology Journal*. 6(2): 30-34.

Ahirrao, R., Pawar, S., Borse, L., Borse, S., Desai, S., Muthu, A. K. 2009. Anthelmintic activity of leaves of *Jatropha curcas* Linn and *Vitex negundo* Linn. *Pharmacology Online News Letters*. 1: 276-279.

Aiyelaagbe, O. O., Adeniyi, B. A., Fatunsin, O. F., Arimah, B. D. 2007. In vitro Antimicrobial Activity and Phytochemical Analysis of *Jatropha curcas* Roots. *International Journal of Pharmacology*. 3(1): 106-110.

Aiyelaagbe, O. O., Hamid, A. A., Fattorusso, E., Taglialatela-Scafati, O., Schrder, H. C., ller, W. E. G. 2011. Cytotoxic Activity of Crude Extracts as well as of Pure Components from *Jatropha* Species, Plants Used Extensively in African Traditional Medicine. *Evidence-Based Complementary and Alternative Medicine*. doi: 10.1155/2011/134954.

Akinpelu, D. A., Aiyegoro, O. A., Okoh, A. I. 2009. The bioactive potentials of two medicinal plants commonly used as folklore remedies among some tribes in West Africa. *African Journal of Biotechnology*. 8(8): 1660-1664.

Aladodo, R. A., Muhammad, N. O., Balogun, E. A. 2013. Effects of aqueous root extract of *Jatropha curcas* on hyperglycaemic and haematological indices in alloxan-induced diabetic rats. *Fountain Journal of Natural and Applied Sciences*. 2(1): 52-58.

Altei, W. F., Picchi, D. G., Abissi, B. M., Giesel, G. M., Flausino Jr, O., Reboud-Ravaux, M., ... Bolzani, V. S. (2014). Jatrophidin I, a cyclic peptide from Brazilian *Jatropha curcas* L.: Isolation, characterization, conformational studies and biological activity. *Phytochemistry*, 107: 91-96.

Alwi, N. A. and Sukardi, S. 2013. Anti implantation effects of *Jatropha curcas* crude oil when fed to pregnant Sprague dawley rats during the early gestation period. *Advances in Life Science and Technology*. 14(1): 1-6.

Ameyaw, E., KK, E. K., Boampong, J., Thomford, A., Anane, R., Tabiri, I., Yaw, A. 2014. Ethanolic Root Extract of *Jatropha curcas* L. Relieves Muscle Pain in Rats. *Research and Reviews: A Journal of Pharmacology*, 4(2): 5-10.

Apu, A. S., Hossain, F., Rizwan, F., Bhuyan, S. H., Matin, M., Jamaluddin, A. 2012. Study of pharmacological activities of methanol extract of Jatropha gossypifolia fruits. *Journal of basic and clinical pharmacy*. 4(1): 20.

Apu, A. S., Ireen, K., Bhuyan, S. H., Matin, M., Hossain, M. F., and Rizwan, F. 2012. Anti-diarrheal Potential of Jatropha gossypifolia (Linn.). *J. Med. Sci.* 12(8): 274-279.

Arekemase, M., Kayode, R., Ajiboye, A. 2011. Antimicrobial activity and phytochemical analysis of *Jatropha curcas* plant against some selected microorganisms. *International Journal of Biology*. 3(3): 52-59.

Asuk, A., Dasofunjo, K., Okafor, A., Mbina, F. 2015. Antidiabetic and antioxidative effects of *Jatropha curcas* extracts in streptozotocin-induced diabetic rats. *British Journal of Medicine and Medical Research*, 5(3): 341-349.

Augustus, G., Jayabalan, M. and Seiler, G. 2002. Evaluation and bioinduction of energy components of *Jatropha curcas*. *Biomass and Bioenergy*. 23(3): 161-164.

Auvin, C., Baraguey, C., Blond, A., Lezenven, F., Pousset, J. L., Bodo, B. 1997. Curcacycline B, a cyclic nonapeptide from *Jatropha curcas* enhancing rotamase activity of cyclophilin. *Tetrahedron Letters*. 38(16): 2845-2848.

Auwal, M., Fatima, A., Shuaibu, A., Ismaila, A., Kyari, A., Thaluvwa, A., Njobdi, A. 2013. Phytochemical properties and effect of aqueous extract of Jatropha caucus root bark on some bacterial isolates. *Journal of Pharmacognosy and Phytotherapy*. 5(8): 148-152.

Azzaz, N. A. E., Rafeek, M. I., El-Sharkawy, Ayman, Y., El-khateeb, H. 2010. Antifungal effect of crude methanolic extract and its fractions of *Jatropha curcas* L., leaves. *J. Biol. Chem*. 5(3): 671-680.

Bailis, R., Kavlak, G. 2013. Environmental implications of Jatropha biofuel from a silvi-pastoral production system in central-west Brazil. *Environ. Sci. Technol*. 47(14): 8042-8050.

Balaji, R., Rekha, N., Deecaraman, M., Manikandan, L. 2009a. Antimetastatic and antiproliferative activity of methanolic fraction of *Jatropha curcas* against B16F10 melanoma induced lung metastasis in C57BL/6 mice. *African Journal of Pharmacy and Pharmacology*. 3(11): 547-555.

Balaji, R., Suba, V., Rekha, N., Deecaraman, M. 2009b. Hepatoprotective activity of methanolic fraction of *Jatropha curcas* on aflatoxin b1 induced hepatic carcinoma. *International Journal of Physical Sciences*, 2, 287-296.

Balangcod, T. D., Balangcod, A. K. D. 2011. Ethnomedical knowledge of plants and healthcare practices among the Kalanguya tribe in Tinoc, Ifugao, Luzon, Philippines. *Indian Journal of Traditional Knowledge*, 10(2): 227-238.

Baumert, S., Khamzina, A., Vlek, P. L. G. 2014. Soil organic carbon sequestration in *Jatropha curcas* systems in Burkina Faso. *Land Degradation and Development*, n/a-n/a. doi: 10. 1002/ldr.2310.

Behera, S. K., Srivastava, P., Tripathi, R., Singh, J., Singh, N. 2010. Evaluation of plant performance of *Jatropha curcas* L. under different agro-practices for optimizing biomass–a case study. *Biomass and Bioenergy*, 34(1): 30-41.

Benge, M. 2006. Assessment of the potential of *Jatropha curcas* (biodiesel tree,) for energy production and other uses in developing countries. *USAID Report*.

Bhagat, R., Ambavade, S., Misar, A., Kulkarni, D. 2011. Anti-inflammatory activity of Jatropha gossypifolia L. leaves in albino mice and Wistar rat. *Journal of Scientific and Industrial Research*, 70(4), 289-292.

Boudjeko, T., Ngomoyogoli, J. E. K., Woguia, A. L., Yanou, N. N. 2013. Partial characterization, antioxidative properties and hypolipidemic effects of oilseed cake of *Allanblackia floribunda* and *Jatropha curcas*. *BMC Complementary and Alternative Medicine*, 13(1): 352.

Brittaine, R., Lutaladio, N. 2010. *Jatropha: a smallholder bioenergy crop: the potential for pro-poor development* (Vol. 8): Food and Agriculture Organization of the United Nations (FAO).

Burkill, H. M. 1995. *The useful plants of west tropical Africa*, Vols. 1-3: Royal Botanic Gardens, Kew.

Camellia, N. N., Lee, A. T., Abdullah, N. 2014. Genetic relationships and diversity of *Jatropha curcas* accessions in Malaysia. *African Journal of Biotechnology*. 11(13): 3048-3054.

Chhabra, S. C., Mahunnah, R. L. A., Mshiu, E. N. 1990. Plants used in traditional medicine in Eastern Tanzania. III. Angiosperms (Euphorbiaceae to Menispermaceae). *Journal of Ethnopharmacology*. 28(3): 255-283.

Chianese, G., Fattorusso, E., Aiyelaagbe, O. O., Luciano, P., Schröder, H. C., Müller, W. E. G., Taglialatela-Scafati, O. 2011. Spirocurcasone, a Diterpenoid with a Novel Carbon Skeleton from *Jatropha curcas*. *Organic Letters*. 13(2): 316-319.

Consoli, R. A., Mendes, N. M., Pereira, J., Santos, B. S., Lamounier, M. A. 1989. Influence of several plant extracts on the oviposition behaviour of Aedes fluviatilis (Lutz) (Diptera: Culicidae) in the laboratory. *Memórias do Instituto Oswaldo Cruz*. 84(1): 47-51.

Cordova-Albores, L. C., Rios, M. Y., Barrera-Necha, L. L., Bautista-Baños, S. 2014. Chemical compounds of a native *Jatropha curcas* seed oil from Mexico and their antifungal effect on *Fusarium oxysporum* f. sp. gladioli. *Industrial Crops and Products*. 62: 166-172.

Dada, E., Ekundayo, F., Makanjuola, O. 2014. Antibacterial activities of *Jatropha curcas* (Linn) on coliforms isolated from surface waters in Akure, Nigeria. *International journal of biomedical science: IJBS*. 10(1): 25.

Dagar, J., Tomar, O., Kumar, Y., Bhagwan, H., Yadav, R., Tyagi, N. 2006. Performance of some under-explored crops under saline irrigation in a semiarid climate in Northwest India. *Land Degradation and Development*. 17(3): 285-299.

Dahake, R., Roy, S., Patil, D., Chowdhary, A., Deshmukh, R. A. 2012. Evaluation of anti-viral activity of *Jatropha curcas* leaf extracts against potentially drug-resistant HIV isolates. *BMC Infectious Diseases*. 12(Suppl. 1): P14.

Daniyan, S. Y., Abalaka, M. E., Elemba, O. M., and Aransiola, S. A. 2011. *In vitro* antimicrobial activity and phytochemical screening of *Jatropha curcas* seed extract. *International research journal of pharmacy*, 2(7): 60-64.

Das, B., Reddy, K., Ravindranath, N., Kishore, K. H., Murthy, U. 2005. A substituted imidazole derivative from *Jatropha curcas*. *Indian Journal of Chemistry B*. 44: 1119-1120.

Demirbas, A., Demirbas, M. F. 2010. *Algae energy: algae as a new source of biodiesel*: Springer Science and Business Media.

Devappa, R. K., Makkar, H. P., Becker, K. 2010. Jatropha toxicity—a review. *Journal of toxicology and environmental health, Part B*. 13(6): 476-507.

Devappa, R. K., Rajesh, S. K., Kumar, V., Makkar, H. P. S., Becker, K. 2012. Activities of *Jatropha curcas* phorbol esters in various bioassays. *Ecotoxicology and Environmental Safety*. 78: 57-62.

Divakara, B., Upadhyaya, H., Wani, S., Gowda, C. L. 2010. Biology and genetic improvement of *Jatropha curcas* L.: a review. *Applied Energy*. 87(3): 732-742.

Edrisi, S. A., Dubey, R. K., Tripathi, V., Bakshi, M., Srivastava, P., Jamil, S., Abhilash, P. 2015. *Jatropha curcas* L.: A crucified plant waiting for resurgence. *Renewable and Sustainable Energy Reviews*. 41: 855-862.

Eguale, T., Giday, M. 2009. In vitro anthelmintic activity of three medicinal plants against *Haemonchus contortus*. *International Journal of Green Pharmacy*. 3(1): 29-34.

Ekundayo, F., Adeboye, C., Ekundayo, E. 2011. Antimicrobial activities and phytochemical screening of pignut (*Jatrophas curcas* Linn.) on some pathogenic bacteria. *Journal of Medicinal Plants Research*. 5(7): 1261-1264.

El-Diwani, G., El-Rafie, S., Hawash, S. 2009. Antioxidant activity of extracts obtained from residues of nodes leaves stem and root of Egyptian *Jatropha curcas*. *African Journal of Pharmacy and Pharmacology*. 3(11): 521-530.

El Diwani, G., Attia, N., Hawash, S. 2009. Development and evaluation of biodiesel fuel and by-products from jatropha oil. *International Journal of Environmental Science and Technology*. 6(2): 219-224.

Esimone, C., Nworu, C., Jackson, C. 2008. Cutaneous wound healing activity of a herbal ointment containing the leaf extract of *Jatropha curcas* L. (Euphorbiaceae). *International Journal of Applied Research in Natural Products*. 1(4): 1-4.

Everson, C. S., Mengistu, M. G., Gush, M. B. 2013. A field assessment of the agronomic performance and water use of *Jatropha curcas* in South Africa. *Biomass and Bioenergy*. 59: 59-69.

FACT. 2007. *Position Paper on Jatropha curcas L. State of the art, small and large scale project development*. Fuels from Agriculture in Communal Technology (available at http://www.fact fuels.org).

Farag, M., Al-Rehaily, A., Ahmad, M. S., Mothana, R. A. 2014. Detection of hypoglycemic and antidiabetic fraction in ethanol extract of *Jatropha curcas* aerial parts. *Pharmacology and Pharmacy*. 5: 663-669.

Feitosa, C., Freitas, R., Luz, N., Bezerra, M., Trevisan, M. T. S. 2011. Acetylcholinesterase inhibition by somes promising Brazilian medicinal plants. *Brazilian Journal of Biology*. 71(3): 783-789.

Félix-Silva, J., Giordani, R. B., Silva-Jr, A. A. d., Zucolotto, S. M., Fernandes-Pedrosa, M. D. F. 2014. Jatropha gossypiifolia L. (Euphorbiaceae): a review of traditional uses, phytochemistry, pharmacology, and toxicology of this medicinal plant. *Evidence-Based Complementary and Alternative Medicine, 2014*.

Foidl, N. and Eder, P. (1997). Agro-industrial exploitation of *J. curcas*. *Biofuels and industrial products from Jatropha curcas*, 88-91.

Gallegos-Tintoré, S., Torres-Fuentes, C., Martínez-Ayala, A. L., Solorza-Feria, J., Alaiz, M., Girón-Calle, J., Vioque, J. 2011. Antioxidant and chelating activity of *Jatropha curcas* L. protein hydrolysates. *Journal of the Science of Food and Agriculture*. 91(9): 1618-1624.

Garg, U. K., Kaur, M., Garg, V., Sud, D. 2007. Removal of hexavalent chromium from aqueous solution by agricultural waste biomass. *Journal of Hazardous Materials*. 140(1): 60-68.

Gbolade, A. A. (2009). Inventory of antidiabetic plants in selected districts of Lagos State, Nigeria. *Journal of Ethnopharmacology*. 121(1): 135-139.

Goel, G., Makkar, H. P., Francis, G., Becker, K. 2007. Phorbol esters: structure, biological activity, and toxicity in animals. *International Journal of Toxicology*. 26(4): 279-288.

Goel, G., Makkar, H. P. S., Francis, G., and Becker, K. 2007. Phorbol Esters: Structure, Biological Activity, and Toxicity in Animals. *International Journal of Toxicology*. 26(4): 279-288.

Goering, C., Daugherty, M., Heakin, A., Pryde, E., Schwab, A. 1982. Fuel properties of eleven vegetable oils. *ASAE Tech. Pap.*; (United States). 25(6): 1472-1483.

Gokhale, D. 2008. Jatropha: Experience of Agro-Forestry and Wasteland Development Foundation, Nashik, India. *International Consultation on Pro-poor Jatropha Development. Rome: Syngenta International.*

Goonasekera, M. M., Gunawardana, V. K., Jayasena, K., Mohammed, S. G., Balasubramaniam, S. 1995. Pregnancy terminating effect of *Jatropha curcas* in rats. *Journal of Ethnopharmacology*. 47(3): 117-123.

Gopinathan, M. C., Sudhakaran, R. 2011. Biofuels: opportunities and challenges in India. *Biofuels* (pp. 173-209): Springer.

Gour, V. 2006. *Production practices including post harvest management of Jatropha curcas.* Paper presented at the Proceedings of the biodiesel conference toward energy independance-Focus of Jatropha, Hyderabad, India.

Grainge, M., Ahmed, S. 1988. *Handbook of plants with pest-control properties*: John Wiley and Sons Limited.

Gübitz, G. M., Mittelbach, M., Trabi, M. 1999a. Exploitation of the tropical oil seed plant *Jatropha curcas* L. *Bioresource Technology*. 67(1): 73-82.

Guedes, R. E., Cruz, F. d. A., Lima, M. C. d., Sant'Ana, L. D. O., Castro, R. N., Mendes, M. F. 2014. Detoxification of *Jatropha curcas* seed cake using chemical treatment: Analysis with a central composite rotatable design. *Industrial Crops and Products*. 52: 537-543.

Gupta, D. D., Haque, M. E., Islam, M. N., Mondal, M. S. I., Shibib, B. A. 2010. Antimicrobial and cytotoxic activities of *Jatropha curcas* (Euphorbiaceae). *Dhaka University Journal of Pharmaceutical Sciences*. 9(2): 139-142.

Haas, W., Sterk, H., Mittelbach, M. 2002. Novel 12-Deoxy-16-hydroxyphorbol diesters isolated from the seed oil of *Jatropha curcas*. *Journal of Natural Products*. 65(10): 1434-1440.

Hamarneh, A., Heeres, H., Broekhuis, A., Picchioni, F. 2010. Extraction of *Jatropha curcas* proteins and application in polyketone-based wood adhesives. *International Journal of Adhesion and Adhesives*. 30(7): 615-625.

Harjono, P., Sugita, M. U. Z. 2012. Synthesis and application of Jatropha oil based polyurethane as paint coating material. *Makara J. Sci*. 16: 134-140.

Heller, J. 1996. Physic nut. *Jatropha curcas* L. Promoting the conservation and use of underutilized and neglected crops. 1. *Roma: IBPGR*.

Henning, R. 2000. *The Jatropha Manual. A guide to the integrated exploitation of the Jatropha plant in Zambia.* Germany: Deutsche Gesellschaft für Technische Zusammenarbeit GTZ.

Henning, R. 2009. The Jatropha system, an integrated approach of rural development. *Gruebels*. 24: 34-80.

Henning, R., Samake, F., Thiero, I. 1995. The nutrient value of Jatropha meal. *Projet Pourghere DNHE-GTZ, Bamako*.

Hirota, M., Suttajit, M., Suguri, H., Endo, Y., Shudo, K., Wongchai, V., Fujiki, H. 1988. A new tumor promoter from the seed oil of *Jatropha curcas* L., an intramolecular diester of 12-deoxy-16-hydroxyphorbol. *Cancer Research*. 48(20): 5800-5804.

Igbinosa, O. O., Igbinosa, I. H., Chigor, V. N., Uzunuigbe, O. E., Oyedemi, S. O., Odjadjare, E. E., Igbinosa, E. O. 2011. Polyphenolic contents and antioxidant potential of stem bark extracts from *Jatropha curcas* (Linn). *International Journal of Molecular Sciences*, 12 (5): 2958-2971.

Igbinosa, O. O., Oviasogie, E. F., Igbinosa, E. O., Igene, O., Igbinosa, I. H., Idemudia, O. G. 2013. Effects of biochemical alteration in animal model after short-term exposure of *Jatropha curcas* (Linn) leaf extract. *The Scientific World Journal, 2013*, 5. doi: 10.1155/2013/798096.

Islam, A. A., Yaakob, Z., Anuar, N., Primandari, S., and Osman, M. (2012). Physiochemical properties of *Jatropha curcas* seed oil from different origins and candidate plus plants (CPPs). *Journal of the American Oil Chemists' Society*. 89(2): 293-300.

Islam, A. A., Yaakob, Z., Ghani, J. A., Anuar, N. 2014. *Jatropha curcas* L.: A Future Energy Crop with Enormous Potential. *Biomass and Bioenergy* (pp. 31-61): Springer.

Jaganath, B., Subramanyam, K., Mayavan, S., Karthik, S., Elayaraja, D., Udayakumar, R., Ganapathi, A. 2014. An efficient in planta transformation of *Jatropha curcas* (L.) and multiplication of transformed plants through in vivo grafting. *Protoplasma*. 251(3): 591-601.

Jain, C., Trivedi, P. 1997. Nematicidal activity of certain plants against root-knot nematode, Meloidogyne incognita infecting chickpea, Cicer arietinum. *Annals of Plant Protection Sciences*. 5(2): 171-174.

Jaiswal, V. 2010. Culture and ethnobotany of Jaintia tribal community of Meghalaya, Northeast India-A mini review. *Indian Journal of Traditional Knowledge* 9(1): 38-44.

Jalalpure, S. S., Alagawadi, K. R., Mahajanshetty, C. S., Salahuddin, M., Shah, B. 2006. In vitro Antihelmintic property of various seed oils. *Iranian Journal of Pharmaceutical Research*. 5(4): 281-284.

Jayakumar, G., Ajithabai, M., Sreedevi, S., Viswanathan, P., Remeshkumar, B. 2010. Ethnobotanical survey of the plants used in the treatment of diabetes. *Indian Journal of Traditional Knowledge*. 9(1): 100-104.

Jing, L., Fang, Y., Ying, X., Wenxing, H., Meng, X., Syed, M. N., Fang, C. 2005. Toxic impact of ingested Jatropherol-I on selected enzymatic activities and the ultrastructure of midgut cells in silkworm, Bombyx mori L. *Journal of Applied Entomology*. 129(2): 98-104.

Johnson, M., Olufunmilayo, L., Adegboyega, C., Adetayo, O. 2014. Evaluation of antidiabetic and the effect of methanolic leaf extract of *Jatropha curcas* on some biochemical parameters in alloxan-induced diabetic male albino rats. *European Journal of Medicinal Plants*. 4(12): 1501-1512.

Jones, M., Csurhes, S. 2008. Pest Plant Risk Assessment Physic nut (*Jatropha curcas*). *Department of Primary Industries and Fisheries, the State of Queensland* 1: 28.

Jongschaap, R., Corré, W., Bindraban, P., Brandenburg, W. 2007. *Claims and Facts on Jatropha curcas L.*

Juan, L., Fang, Y., Lin, T., Fang, C. 2002. Isolation, Purification and Functional Investigation on the N-glycosidase Activity of Curcin from the Seeds of *Jatropha curcas. High Technology Letters, 11*.

Kalam, M., Masjuki, H., Shahabuddin, M., Mofijur, M. 2012. Tribological characteristics of amine phosphate and octylated/butylated diphenylamine additives infused bio-lubricant. *Energy Education Science and Technology*. 30: 123-136.

Kalimuthu, K., Vijayakumar, S., Senthilkumar, R. 2010. Antimicrobial activity of the biodiesel plant, *Jatropha curcas* L. *International Journal of Pharma and Bio Sciences*, 1 (3): 1-5.

Kant, P., Wu, S. 2011. The extraordinary collapse of Jatropha as a global biofuel. *Environ. Sci. Technol.* 45(17): 7114-7115.

Karmee, S. K., Chadha, A. 2005. Preparation of biodiesel from crude oil of Pongamia pinnata. *Bioresource Technology*. 96(13): 1425-1429.

Karthikeyan, S., Sivakumar, P., Palanisamy, P. 2008. Novel activated carbons from agricultural wastes and their characterization. *Journal of Chemistry*. 5(2): 409-426.

Kaushik, N., Kumar, S. 2005. *Jatropha curcas L.: silviculture and uses*: Agrobios.

Ke, C., Peng, R., ChunYin, Y., Qi, J., XiuQing, J. 2011. Genetic relationships among *Jatropha curcas* L. clones from Panzhihua, China as revealed by RAPD and ISSR. *African Journal of Agricultural Research*. 6 11: 2582-2585.

Khafagy, S., Mohamed, Y., Salam, N. A., Mahmoud, Z. 1977. Phytochemical study of *Jatropha curcas*. *Planta Medica*. 31(3): 274-277.

Kharat, A., Dolui, A., Das, S. 2011. Free radical scavenging potential of Jatropha gossypifolia. *Asian Journal of Chemistry*. 23(2): 799-801.

Kim, M. J., Yang, S. W., Mao, H. Z., Veena, S. P., Yin, J. L., Chua, N. H. 2014. Gene silencing of Sugar-dependent 1 (JcSDP1), encoding a patatin-domain triacylglycerol lipase, enhances seed oil accumulation in *Jatropha curcas*. *Biotechnology for Biofuels*. 7 (1): 36.

King, A. J., He, W., Cuevas, J. A., Freudenberger, M., Ramiaramanana, D., and Graham, I. A. (2009). Potential of *Jatropha curcas* as a source of renewable oil and animal feed. *Journal of Experimental Botany*. 60(10): 2897-2905.

Kong, L. Y., Min, Z. D., Shi, J. X., Rui, F. 1996. Chemical constituents from roots of *Jatropha curcas*. *Acta Botanica Sinica-Chinese Edition*. 38: 161-166.

Kumar, A., Sharma, S. 2008. An evaluation of multipurpose oil seed crop for industrial uses (*Jatropha curcas* L.): a review. *Industrial Crops and Products*. 28(1): 1-10.

Kumar, M. S., Yaakob, Z., Maimunah, S., Abdullah, S. 2010. Synthesis of alkyd resin from non-edible Jatropha seed oil. *Journal of Polymers and the Environment*. 18(4): 539-544.

Kumar, S., Gupta, A. and Naik, S. 2003. Conversion of non-edible oil into biodiesel. *Journal of Scientific and Industrial Research*. 62(1/2): 124-132.

Kumari, V., Gupta, S. and Roy, S. 2014. Study of phytochemical analysis and biological activities of *Jatropha curcas* L(Euphorbiaceae). *World Journal of Pharmacy and Pharmaceutical Sciences*. 3(9): 959-969.

Lestari, D., Mulder, W. J., Sanders, J. P. 2011. Jatropha seed protein functional properties for technical applications. *Biochemical Engineering Journal*. 53(3): 297-304.

Li, X., Li, L., Wang, J., Wang, T., Wang, L. 2014. Two new lignans with antioxidative activities from *Jatropha curcas*. *Natural Product Research*. 28(22): 1985-1991.

Lin, J., Chen, Y., Xu, Y., Yan, F., Tang, L., Chen, F. 2003a. Cloning and expression of curcin, a ribosome-inactivating protein from the seeds of *Jatropha curcas*. *Acta Botanica Sinica*. 45(7): 858-863.

Lin, J., Yan, F., Tang, L., Chen, F. 2003b. Antitumor effects of curcin from seeds of *Jatropha curcas*. *Acta Pharmacologica Sinica*. 24(3): 241-246.

Lin, J., Zhou, X., Wang, J., Jiang, P., Tang, K. 2010. Purification and characterization of Curcin, a toxic lectin from the seed of *Jatropha curcas*. *Preparative Biochemistry and Biotechnology*. 40(2): 107-118.

Liu, J. Q., Yang, Y. F., Li, X. Y., Liu, E. Q., Li, Z.-R., Zhou, L., Qiu, M. H. 2013. Cytotoxicity of naturally occurring rhamnofolane diterpenes from *Jatropha curcas*. *Phytochemistry*. 96: 265-272.

Liu, J. Q., Yang, Y. F., Wang, C. F., Li, Y., Qiu, M. H. 2012. Three new diterpenes from *Jatropha curcas*. *Tetrahedron*. 68(4): 972-976.

Lopes, S. L., Fragoso, R., Duarte, E., Marques, P. A. S. S. 2015. Bioconversion of *Jatropha curcas* seed cake to hydrogen by a strain of Enterobacter aerogenes. *Fuel*. 139: 715-719.

Luo, M. J., Liu, W. X., Yang, X. Y., Xu, Y., Yan, F., Huang, P., Chen, F. 2007. Cloning, expression, and antitumor activity of recombinant protein of curcin. *Russian Journal of Plant Physiology*. 54(2): 202-206.

Luo, M. J., Yang, X. Y., Liu, W. X., Ying, X. U., Huang, P., Yan, F., Chen, F. 2006. Expression, purification and anti-tumor activity of curcin. *Acta Biochimica et Biophysica Sinica*. 38(9): 663-668.

Maciel, F. M., Laberty, M. A., Oliveira, N. D., Felix, S. P., Soares, A. M. d. S., Verícimo, M. A., Machado, O. L. T. 2009. A new 2S albumin from *Jatropha curcas* L. seeds and assessment of its allergenic properties. *Peptides*. 30(12): 2103-2107.

Makkar, H., Becker, K., Schmook, B. 1998. Edible provenances of *Jatropha curcas* from Quintana Roo state of Mexico and effect of roasting on antinutrient and toxic factors in seeds. *Plant Foods for Human Nutrition*. 52(1): 31-36.

Makkar, H. P., Francis, G. and Becker, K. 2008. Protein concentrate from *Jatropha curcas* screw-pressed seed cake and toxic and antinutritional factors in protein concentrate. *Journal of the Science of Food and Agriculture*. 88(9): 1542-1548.

Makonnen, E., Rostom, A. A., Assefa, G., Zerihun, L. 1997. Antifertility effect of *Jatropha curcas* L. seed in guinea pigs. *Ethiop. J. Health Dev*. 11(2): 145-148.

Maravi, D. K., Mazumdar, P., Alam, S., Goud, V. V., and Sahoo, L. 2015. Jatropha (*Jatropha curcas* L.) *Agrobacterium Protocols* (pp. 25-35): Springer.

Matsuse, I. T., Lim, Y. A., Hattori, M., Correa, M., Gupta, M. P. 1998. A search for anti-viral properties in Panamanian medicinal plants.: The effects on HIV and its essential enzymes. *Journal of Ethnopharmacology* 64(1): 15-22.

Meshram, P., Kulkarni, N., Joshi, K. 1994. Antifeedant activity of certain plant products against teak skeletonizer, Eutectona machaeralis Walk. (Lepidoptera: Pyralidae). *Annals of Entomology*. 12(2): 53-56.

Messemaker, L. (2008). *Assessment of the Jatropha value chain and its potential for pro poor biofuel development in Northern Tanzania.* [MSc thesis] International development studies at the Faculty of Geosciences Utrecht University.

Mishra, S. B., Vijayakumar, M., Ojha, S. K., Verma, A. 2010. Antidiabetic effect of *Jatropha curcas* L. leaves extract in normal and alloxan-induced diabetic rats. *Int. J. Ph. Sci*. 2(1): 482-487.

Mitra, C., Bhatnagar, S., Sinha, M. 1970. Chemical examination of *Jatropha curcas*. *Indian Journal of Chemistry*. 8: 1047.

Mofijur, M., Masjuki, H., Kalam, M., Shahabuddin, M., Hazrat, M., Liaquat, A. 2012. Palm oil methyl ester and its emulsions effect on lubricant performance and engine components wear. *Energy Procedia*. 14: 1748-1753.

Monteiro, M. V., Bevilaqua, C. M., Morais, S. M., Machado, L. K., Camurca-Vasconcelos, A. L., Campello, C. C., Mesquita Mde, A. 2011. Anthelmintic activity of *Jatropha curcas* L. seeds on *Haemonchus contortus*. *Vet. Parasitol.* 182(2-4): 259-263.

Muangman, S., Thippornwong, M., Tohtong, R. 2005. Anti-metastatic effects of curcusone B, a diterpene from *Jatropha curcas*. *In Vivo.* 19(1): 265-268.

Muanza, D. N., Euler, K. L., Williams, L., Newman, D. J. 1995. Screening for Antitumor and Anti-HIV Activities of Nine Medicinal Plants from Zaire. *Pharmaceutical Biology*, 33 (2): 98-106.

Mujumdar, A. M., Misar, A. V. 2004. Anti-inflammatory activity of *Jatropha curcas* roots in mice and rats. *Journal of Ethnopharmacology*. 90(1): 11-15.

Murali Krishna, M. V., Seshagiri Rao, V. V., Murthy, P. V. 2014. Performance evaluation of medium grade low heat rejection diesel engine with carbureted methanol and crude jatropha oil. *Renewable and Sustainable Energy Reviews*. 34(C): 122-135.

Naengchomnong, W., Tarnchompoo, B., Thebtaranonth, Y. 1994. (+)-Jatrophol, (+)-Marmesin, Propacin and Jatrophin from the roots of *Jatropha curcas* (Euphorbiaceae). *Journal of the Science Society of Thailand*. 20(2): 73-83.

Naengchomnong, W., Thebtaranonth, Y., Wiriyachitra, P., Okamoto, K. T., and Clardy, J. (1986a). Isolation and structure determination of four novel diterpenes from *Jatropha curcus*. *Tetrahedron Letters*. 27(22): 2439-2442.

Naengchomnong, W., Thebtaranonth, Y., Wiriyachitra, P., Okamoto, K. T., Clardy, J. 1986b. Isolation and structure determination of two novel Lathyrenes from Jatropha Curcus. *Tetrahedron Letters*. 27(47): 5675-5678.

Nagendramma, P. and Kaul, S. (2012). Development of ecofriendly/biodegradable lubricants: An overview. *Renewable and Sustainable Energy Reviews*. 16(1): 764-774.

Nakazibwe, S., Taher, M., Kumar Saxena, A., Ichwan, A., Jauhari, S. 2011. Antiproliferative activity of curcusone B from *Jatropha curcas* on human cancer cell lines. *Australian Journal of Basic and Applied Sciences*. 5(8): 47-51.

Namuli, A., Abdullah, N., Sieo, C., Zuhainis, S., Oskoueian, E. 2011. Phytochemical compounds and antibacterial activity of *Jatropha curcas* Linn. extracts. *Journal of Medicinal Plant Research.* 5: 3982-3990.

Nath, L. K., Dutta, S. K. 1991. Extraction and purification of curcain, a protease from the Latex of *Jatropha curcas* Linn. *Journal of Pharmacy and Pharmacology*. 43(2): 111-114.

Nayak, B., Patel, K. 2010a. Anti-inflammatory screening of *Jatropha curcas* root, stem and leaf in albino rats. *Rom. J. Biol. Plant Biol.* 55: 9-13.

Nayak, B. S., Patel, K. N. 2010b. Screening of different parts of *Jatropha curcas* for antinociceptive and antipyretic activity on rats. *Malaysian Journal of Pharmaceutical Sciences*. 8: 23-28.

Nazir, N., Mangunwidjaja, D., Setyaningsih, D., Yuliani, S., Yarmo, M. A., Salimon, J., Ramli, N. 2014. In-situ Alkaline Transesterification of *Jatropha curcas* seed Oil for Production of Biodiesel and Nontoxic Jatropha seed Cake. *International Journal on Advanced Science, Engineering and Information Technology*. 4(1): 07-13.

Neuwinger, H. D. 1994. *African medicinal plants and poisonous plants used for hunting*: WV GesmbH, Germany.

Neuwinger, H. D. 1996. *African ethnobotany: poisons and drugs: chemistry, pharmacology, toxicology*: CRC Press.

Nordqvist, P., Thedjil, D., Khosravi, S., Lawther, M., Malmström, E., Khabbaz, F. 2012. Wheat gluten fractions as wood adhesives—glutenins versus gliadins. *Journal of Applied Polymer Science*. 123(3): 1530-1538.

Obasi, N., Ejikeme, M., Egbuonu, C. 2011. Antimicrobial and phytochemical activity of methanolic extract and its fractions of *Jatropha curcas* Linn. (Euphorbiaceae) stem bark. *African Journal of Pure and Applied Chemistry*. 5(5): 92-96.

Oduola, T., Adeosun, G. O., Oduola, T. A., Avwioro, G. O., Oyeniyi, M. A. 2005. Mechanism of action of Jatropha gossypifolia stem latex as a haemostatic agent.

Odusote, M., Abioye, A., Coker, H. 1999. The latex of *Jatropha curcas* linn (Euphorblaceae): A prospective haemostatic agent. *Nigerian Quarterly Journal of Hospital Medicine*. 9(2): 158-166.

Odusote, O., Abioye, A., Rotibi, M. 2002. *Jatropha curcas* seed oil linn (euphor biaceae): Contraceptive activity and an oral formulation. *Nigerian Quarterly Journal of Hospital Medicine*. 12(1): 44-47.

Okieimen, F., Sogbaike, C. 1996. Stabilising effect of derivatives of Jatropha seed oil on the thermal degradation of poly (vinyl chloride). *European Polymer Journal*. 32(12): 1457-1462.

Okuda, T. 1995. [Tannins, a new family of bio-active natural organic compounds (questions and answers)]. *Yakugaku zasshi: Journal of the Pharmaceutical Society of Japan*. 115(2): 81-100.

Oliver-Bever, B. 1986. *Medicinal plants in tropical West Africa*: Cambridge University Press.

Oloyede, O., Salau, A., Akeusola, R., Ganiyu, O., Azeez, L., Ogunbode, S. 2012. Phytochemical content, radical scavenging and antibacterial properties of aqueous extract of *Jatropha curcas* Linn leaves. *Fountain Journal of Natural and Applied Sciences*. 1(1): 41-48.

Olukunle, J. O., Adenubi, O. T., Oladele, G. M., Sogebi, E. A., Oguntoke, P. C. 2011. Studies on the anti-inflammatory and analgesic properties of *Jatropha curcas* leaf extract. *Acta Veterinaria Brno*. 80(3): 259-262.

Omeh, Y. S., Ezeja, M. I. 2013. Analgesic activity of the methanolic leaf extract of *Jatropha curcas* (Linn). *African Journal of Biomedical Research*. 13(2): 149-152.

Ong, H. C., Mahlia, T. M. I., Masjuki, H. H., Norhasyima, R. S. 2011. Comparison of palm oil, *Jatropha curcas* and Calophyllum inophyllum for biodiesel: A review. *Renewable and Sustainable Energy Reviews*. 15(8): 3501-3515.

Openshaw, K. 2000. A review of *Jatropha curcas*: an oil plant of unfulfilled promise. *Biomass and Bioenergy*. 19(1): 1-15.

Oskoueian, E., Abdullah, N. and Ahmad, S. (2012). Phorbol esters from *Jatropha* meal triggered apoptosis, activated PKC-δ, caspase-3 proteins and down-regulated the proto-oncogenes in MCF-7 and HeLa cancer cell lines. *Molecules*. 17(9): 10815-10830.

Oskoueian, E., Abdullah, N., Ahmad, S., Saad, W. Z., Omar, A. R., Ho, Y. W. 2011b. Bioactive compounds and biological activities of *Jatropha curcas* L. kernel meal extract. *International journal of molecular sciences*. 12(9): 5955-5970.

Oskoueian, E., Abdullah, N., Saad, W. Z., Omar, A. R., Ahmad, S., Kuan, W. B., Ho, Y. W. 2011a. Antioxidant, anti-inflammatory and anticancer activities of methanolic extracts from *Jatropha curcas* Linn. *J. Med. Plants Res*. 5(1): 49-57.

Osoniyi, O., Onajobi, F. 2003. Coagulant and anticoagulant activities in *Jatropha curcas* latex. *Journal of Ethnopharmacology*. 89(1): 101-105.

Othman, A. R., Abdullah, N., Ahmad, S., Ismail, I. S., Zakaria, M. P. 2015. Elucidation of in-vitro anti-inflammatory bioactive compounds isolated from *Jatropha curcas* L. plant root. *BMC Complementary and Alternative Medicine*. 15(1): 11.

Ouwens, D., Wani, S., Sreedevi, T., Marimuthu, S. 2008. Pro-poor biodiesel initiative for rehabilitating degraded drylands. *International Crops Research Institute for the Semi-Arid Tropics (ICRISAT) In International Consultation on Pro-poor Jatropha Development*. 10-11.

Oyi, A., Onaolapo, J., Haruna, A., Morah, C. 2007. Antimicrobial screening and stability studies of the crude extract of *Jatropha curcas* Linn latex (Euphorbiaceae). *Nigerian Journal of Pharmaceutical Sciences*. 6(2): 14-20.

Palash, S. M., Kalam, M. A., Masjuki, H. H., Arbab, M. I., Masum, B. M., Sanjid, A. 2014. Impacts of NOx reducing antioxidant additive on performance and emissions of a multi-cylinder diesel engine fueled with Jatropha biodiesel blends. *Energy Conversion and Management*. 77: 577-585.

Patade, V. Y., Khatri, D., Kumar, K., Grover, A., Kumari, M., Gupta, S. M., Nasim, M. 2014. RNAi Mediated curcin precursor gene silencing in Jatropha (*Jatropha curcas* L.). *Molecular Biology Reports*. 41(7): 4305-4312.

Patel, V., Varughese, J., Krishnamoorthy, P., Jain, R., Singh, A., Ramamoorty, M. 2008. Synthesis of alkyd resin from Jatropha and rapeseed oils and their applications in electrical insulation. *Journal of Applied Polymer Science*. 107(3): 1724-1729.

Patil, D., Roy, S., Dahake, R., Rajopadhye, S., Kothari, S., Deshmukh, R., Chowdhary, A. 2013. Evaluation of *Jatropha curcas* Linn. Leaf extracts for its cytotoxicity and potential to inhibit hemagglutinin protein of influenza virus. *Indian Journal of Virology*. 24(2): 220-226.

Pertino, M., Schmeda-Hirschmann, G., Rodríguez, J. A., Theoduloz, C. 2007. Gastroprotective effect and cytotoxicity of semisynthetic jatropholone derivatives. *Planta Medica*. 73(10): 1095-1100.

Pratt, J., Henry, E., Mbeza, H., Mlaka, E., Satali, L. 2002. Malawi Agroforestry Extension Project Marketing and Enterprise Program, Main Report. *Malawi Agroforestry*. 139.

Pryor, R., Hanna, M., Schinstock, J., Bashford, L. 1983. Soybean oil fuel in a small diesel engine. *Transactions of the ASAE [American Society of Agricultural Engineers]*.

Rachana, S., Tarun, A., Rinki, R., Neha, A., Meghna, R. 2012. Comparative analysis of antibacterial activity of *Jatropha curcas* fruit parts. *Journal of Pharmaceutical and Biomedical Sciences (JPBMS)*. 15(15): 1-4.

Rahman, M., Ahmad, S. H., Mohamed, M. T. M., Rahman, M. Z. A. 2013. Extraction of *Jatropha curcas* fruits for antifungal activity against anthracnose (*Colletotrichum gloeosporioides*) of papaya. *African Journal of Biotechnology*. 10(48): 9796-9799.

Ramakrishnan, K., Namasivayam, C. 2009. Development and characteristic of activated carbons from Jatropha husk, an agro industrial solid waste, by chemical activation methods. *Journal Environment Engineering Management*. 9: 173-178.

Ramesan, M. 2005. The effects of filler content on cure and mechanical properties of dichlorocarbene modified styrene butadiene rubber/carbon black composites. *Journal of Polymer Research*. 11(4): 333-340.

Ravindranath, N., Ramesh, C., Das, B. 2003. A rare dinorditerpene from *Jatropha curcas*. *Biochemical systematics and ecology*. 31(4): 431-432.

Ravindranath, N., Ravinder Reddy, M., Mahender, G., Ramu, R., Ravi Kumar, K., Das, B. 2004a. Deoxypreussomerins from *Jatropha curcas*: are they also plant metabolites? *Phytochemistry*. 65(16): 2387-2390.

Ravindranath, N., Ravinder Reddy, M., Ramesh, C., Ramu, R., Prabhakar, A., Jagadeesh, B., Das, B. 2004b. New lathyrane and podocarpane diterpenoids from *Jatropha curcas*. *Chemical and Pharmaceutical Bulletin*. 52(5): 608-611.

Ravindranath, N., Venkataiah, B., Ramesh, C., Jayaprakash, P., Das, B. 2003. Jatrophenone, a novel macrocyclic bioactive diterpene from *Jatropha gossypifolia*. *Chemical and Pharmaceutical Bulletin*. 51(7): 870-871.

Reena, P. (2011). Evaluation of antimicrobial and anti-inflammatory activities of bark of Jatropha gossypifolia. *World Journal of Science and Technology*. 1(10): 00-00.

Reyadh, M. 2004. The cultivation of *Jatropha curcas* in Egypt. Under Secretary of State for Afforestation Ministry of Agriculture and Land Reclamation, (Retrieve from http://www.fao.org).

Roach, J. S., Devappa, R. K., Makkar, H. P., Becker, K. 2012. Isolation, stability and bioactivity of *Jatropha curcas* phorbol esters. *Fitoterapia*. 83(3): 586-592.

Rosado, T. B., Laviola, B. G., Faria, D. A., Pappas, M. R., Bhering, L. L., Quirino, B., Grattapaglia, D. 2010. Molecular Markers Reveal Limited Genetic Diversity in a Large Germplasm Collection of the Biofuel Crop L. in Brazil. *Crop Science*. 50(6): 2372-2382.

Sachdeva, K., Garg, P., Singhal, M., Srivastava, B. 2011. Wound healing potential of extract of *Jatropha curcas* l. (stem bark) in rats. *Pharmacognosy Journal*. 3(25): 67-72.

Saetae, D., Suntornsuk, W. 2010. Antifungal activities of ethanolic extract from *Jatropha curcas* seed cake. *Journal of Microbiology and Biotechnology*. 20(2): 319-324.

Salas, J., Tello, V., Zavaleta, A., Villegas, L., Salas, M., Fernández, I., Vaisberg, A. 1994. Cicatrization effect of *Jatropha curcas* latex (Angiospermae: Euphorbiaceae). *Revista de Biologia Tropical*. 42(1-2): 323-326.

Salih, N., Salimon, J., Yousif, E. 2012. Synthetic biolubricant basestocks based on environmentally friendly raw materials. *Journal of King Saud University-Science*. 24(3): 221-226.

Sangeetha, J., Divya, K., Vamsikrishna, M. P. A., Rani, G. L. 2010. Anti-inflammatory and antibacterial activity of *Jatropha curcas* linn. *Asian Journal of Pharmaceutical Research and Health Care*. 2(3): 258-262.

Sathishkumar, P., Arulkumar, M., Palvannan, T. 2012. Utilization of agro-industrial waste *Jatropha curcas* pods as an activated carbon for the adsorption of reactive dye Remazol Brilliant Blue R (RBBR). *Journal of Cleaner Production*. 22(1): 67-75.

Satyawan, D., Tasma, I. M. 2011. Genetic Diversity Analysis of *Jatropha curcas* Provenances Using Randomly Amplified Polymorphic DNA Markers. *Jurnal AgroBiogen*. 7(1): 47-55.

Setha, B., Laga, A., Mahendradatta, M. (2014). Phytochemical screening and antibacterial activity of crude extracts from *Jatropha curcas*, Linn against histamine-forming bacteria. *Elixir Org. Chem*. 67: 21438-21441.

Shahabuddin, M., Masjuki, H., Kalam, M. 2013. Experimental investigation into tribological characteristics of bio-lubricant formulated from Jatropha oil. *Procedia Engineering*. 56: 597-606.

Shahwar, D., Ahmad, N., Ullah, S., Raza, M. A. 2010. Antioxidant activities of the selected plants from the family Euphorbiaceae, Lauraceae, Malvaceae and Balsaminaceae. *African Journal of Biotechnology*. 9(7): 1086-1096.

Sharma, A. K., Gangwar, M., Tilak, R., Nath, G., Sinha, A. S. K., Tripathi, Y. B., Kumar, D. 2012. Comparative in vitro antimicrobial and phytochemical evaluation of methanolic extract of root, stem and leaf of *Jatropha curcas* Linn. *Pharmacognosy Journal*. 4(30): 34-40.

Sharma, J., Painuli, R., Gaur, R. 2010. Plants used by the rural communities of district Shahjahanpur, Uttar Pradesh. *Indian Journal of Traditional Knowledge*. 9(4): 798-803.

Shetty, S., Udupa, S. L., Udupa, A. L., Vollala, V. R. 2006. Wound healing activities of Bark Extract of *Jatropha curcas* Linn in albino rats. *Saudi Medical Journal*, 27(10): 1473-1476.

Shivani, P., Khushbu, P., Faldu, N., Thakkar, V., Shubramanian, R. 2013. Extraction and analysis of *Jatropha curcas* L. seed oil. *African Journal of Biotechnology*. 10(79); 18210-18213.

Silva, S. D. N., Abreu, I. C., Freire, S. M. D. F., Cartágenes, M. D. S. D. S., Ribeiro, R. M., Castro, A. S. D., Borges, M. O. D. R. 2011. Antispasmodic effect of Jatropha gossypiifolia is mediated through dual blockade of muscarinic receptors and Ca2+ channels. *Revista Brasileira de Farmacognosia*. 21(4): 715-720.

Singh, B., Singh, K., Shukla, G., Goel, V., Pathre, U. V., Rahi, T., Tuli, R. 2013. The field performance of some accessions of *Jatropha curcas* L. (Biodiesel Plant) on degraded sodic land in North India. *International Journal of Green Energy*. 10(10): 1026-1040.

Singh, D., Singh, A. 2005. The toxicity of four native Indian plants: Effect on AChE and acid/alkaline phosphatase level in fish *Channa marulius*. *Chemosphere*. 60(1): 135-140.

Singh, K., Singh, B., Verma, S. K., Patra, D. 2014. *Jatropha curcas*: A ten year story from hope to despair. *Renewable and Sustainable Energy Reviews*. 35: 356-360.

Singh, R., Vyas, D., Srivastava, N., Narra, M. 2008. SPRERI experience on holistic approach to utilize all parts of *Jatropha curcas* fruit for energy. *Renewable Energy*. 33(8): 1868-1873.

Siniawski, M. T., Saniei, N., Adhikari, B., Doezema, L. A. 2007. Influence of fatty acid composition on the tribological performance of two vegetable-based lubricants. *Journal of Synthetic Lubrication*. 24(2): 101-110.

Sorate, K. A., Bhale, P. V. 2015. Biodiesel properties and automotive system compatibility issues. *Renewable and Sustainable Energy Reviews*. 41: 777-798.

Sotolongo, J. A., Beatón, P., Díaz, A., de Oca, S. M., del Valle, Y., Pavón, S. G., Zanzi, R. 2009. *Jatropha curcas* L. as a source for the production of biodiesel: A Cuban experience. *Available: hem. fyristorg. com/zanzi/paper W, 2257.*

Sriprang, S., Sriprang, N., Sumpradit, T., Shimbhu, D. (2010). Antibacterial activities of crude extracts from physic nut (*Jatropha curcas*) seed residues. *Sci. Asia*. 36: 346-348.

Srivastava, S. K., Tewari, J. and Shukla, D. (2008). A folk dye from leaves and stem of *Jatropha curcas* L. used by Tharu tribes of Devipatan division. *Indian Journal of Traditional Knowledge*. 7(1): 77-78.

Staubmann, R., Schubert-Zsilavecz, M., Hiermann, A., Karting, T. 1997. The anti-inflammatory effect of *J. curcas* leaves. G.M. Gübitz, M. Mittelbach, M. Trabi (Eds.), *Biofuels and Industrial Products from Jatropha curcas, Dbv-Verlag Für die Technische Universität Graz,* Graz, Austria, 60-87.

Staubmann, R., Schubert-Zsilavecz, M., Hiermann, A., Kartnig, T. 1999b. A complex of 5-hydroxypyrrolidin-2-one and pyrimidine-2,4-dione isolated from *Jatropha curcas*. *Phytochemistry*. 50(2): 337-338.

Stripe, F., Pession-Brizzi, A., Lorenzoni, E. 1976. Studies on the proteins from the seeds of *Croton toglium* and of *Jatropha curcas*. *Biochemical Journal*. 156: 1-6.

Subroto, E., Manurung, R., Heeres, H. J., Broekhuis, A. A. 2015. Optimization of mechanical oil extraction from *Jatropha curcas* L. kernel using response surface method. *Industrial Crops and Products*. 63: 294-302.

Suhaili, Z., Yeo, C. C., Yasin, H., Badaludin, N., Zakaria, Z. A. 2011. Antibacterial profile of *Jatropha curcas* latex extracts against selected human pathogenic bacteria. *African Journal of Microbiology Research*. 5(29): 5147-5154.

Sun, Q. B., Li, L. F., Li, Y., Wu, G. J., Ge, X. J. 2008. SSR and AFLP markers reveal low genetic diversity in the biofuel plant in China. *Crop Science*. 48(5): 1865-1871.

Sundar, V. J., Muralidharan, C., Mandal, A. B. 2013. Eco-benign stabilization of skin protein—role of *Jatropha curcas* oil as a co-tanning agent. *Industrial Crops and Products*. 47: 227-231.

Sutthivaiyakit, S., Mongkolvisut, W., Prabpai, S., and Kongsaeree, P. 2009. Diterpenes, Sesquiterpenes, and a Sesquiterpene–Coumarin Conjugate from Jatropha integerrima. *Journal of Natural Products*. 72(11): 2024-2027.

Tanya, P., Taeprayoon, P., Hadkam, Y., Srinives, P. 2011. Genetic diversity among Jatropha and Jatropha-related species based on ISSR markers. *Plant Molecular Biology Reporter*. 29(1): 252-264.

Tao, Y.-B., Luo, L., He, L.-L., Ni, J., Xu, Z.-F. 2014. A promoter analysis of Mother of FT and TFL1 1 (JcMFT1), a seed-preferential gene from the biofuel plant *Jatropha curcas*. *Journal of plant research*, 1-12.

Theoduloz, C., Rodríguez, J. A., Pertino, M., and Schmeda-Hirschmann, G. (2009). Antiproliferative activity of the diterpenes jatrophone and jatropholone and their derivatives. *Planta Medica*. 75n(14): 1520-1522. doi: 10.1055/s-0029-1185834.

Thippornwong, M., Tohtong, R. 2006. Effect of curcusone B on KKU-100 cell invasion. *Cancer Epidemiology Biomarkers and Prevention*. 15(12 Supplement): A112.

Thuneke, K., Wilharm, T., Stotz, K. 2005. *Untersuchung der Wechselwirkungen zwischen Rapsöl als Kraftstoff und dem Motorenöl in pflanzenöltauglichen Motoren*: Technologie- und Förderzentrum im Kompetenzzentrum für Nachwachsende Rohstoffe.

Tikkoo, A., Yadav, S. S., Kaushik, N. 2013. Effect of irrigation, nitrogen and potassium on seed yield and oil content of *Jatropha curcas* in coarse textured soils of northwest India. *Soil and Tillage Research*. 134: 142-146.

Timothy, A. W., Maslam, J., Guo, X. A., Mihail, I., Javni, I. J., Zoran, S. P. (2007) WO2007127379.

Uche, F., Aprioku, J. 2008. The phytochemical constituents, analgesic and anti-inflammatory effects of methanol extract of *Jatropha curcas* leaves in Mice and Wister albino rats. *Journal of Applied Sciences and Environmental Management*. 12(4), 99-102.

Van den Berg, A. J. J., Horsten, S. F. A. J., Kettenes-van den Bosch, J. J., Kroes, B. H., Beukelman, C. J., Leeflang, B. R., Labadie, R. P. 1995. Curcacycline A - a novel cyclic octapeptide isolated from the latex of *Jatropha curcas* L. *FEBS Letters*. 358(3): 215-218.

Vaz, C. M., de Graaf, L. A., Mulder, W. J. 2003. Adhesives coatings and bioplastics from protein sources. In: Fahnestock, S. R., Steinbuchel, A., editors. *Polyamides and complex proteinaceous materials II*. New York: Wiley-VCH; p. 383-404.

Villaseñor, I. M., Cariño, F. A. 2011. Antimicrobial activity of new phorbins from *Jatropha curcas* Linn. (Euphorbiaceae) leaves. *Zeitschrift fur Naturforschung C-Journal of Biosciences*. 66(9): 441.

Villegas, L. F., Fernández, I. D., Maldonado, H., Torres, R., Zavaleta, A., Vaisberg, A. J., Hammond, G. B. 1997. Evaluation of the wound-healing activity of selected traditional medicinal plants from Perú. *Journal of Ethnopharmacology*. 55(3): 193-200.

Wang, X. C., Zheng, Z. P., Gan, X. W., Hu, L. H. 2009. Jatrophalactam, a novel diterpenoid lactam isolated from *Jatropha curcas*. *Organic Letters*. 11(23): 5522-5524.

Watt, J. M., Breyer-Brandwyk, G. 1962. *The medicinal and poisonous plants of southern and eastern Africa*. E. and S. Livingstone Ltd., Edinburg and London.

Wei, Q., Liao, Y., Chen, Y., Wang, S., Xu, Y., Tang, L., Chen, F. 2005. Isolation, characterisation and antifungal activity of beta-1, 3-glucanase from seeds of *Jatropha curcas*. *South African Journal of Botany*. 71(1): 95-99.

Wei, Q., Liao, Y., Zhou, L., Zhou, J., Wang, S., Chen, F. 2004. Antifugal activity of curcin from seeds of *Jatropha curcas*. *Chin. J. Oil Crop Sci.* 26(3): 71-75.

Wender, P. A., Kee, J.-M. and Warrington, J. M. 2008. Practical synthesis of prostratin, DPP, and their analogs, adjuvant leads against latent HIV. *Science*. 320(5876): 649-652.

Winkler, E., Foidl, N., Gübitz, G. M., Staubmann, R., Steiner, W. 1997. Enzyme-supported oil extraction from *Jatropha curcas* seeds. *Biotechnology for Fuels and Chemicals*, Proceedings of the Eighteenth symposium on Biotechnology for fuels and chemicals, held at Gatlinburg, Tennessee, US, 5-9 May, pp. 449-456.

Xiao, J., Zhang, H., Niu, L., Wang, X. 2011. Efficient screening of a novel antimicrobial peptide from *Jatropha curcas* by cell membrane affinity chromatography. *Journal of Agricultural and Food Chemistry*. 59(4): 1145-1151.

Xu, J. J., Fan, J. T., Zeng, G. Z., Tan, N. H. 2011. A new tetracyclic diterpene from *Jatropha curcas*. *Helvetica Chimica Acta*. 94(5): 842-846.

Yadav, N., Kumar, R., Rawat, L., Gupta, S. 2014. Physico-Chemical Properties of Before and After Anaerobic Digestion of Jatropha Seed Cake and Mixed With Pure Cow Dung. *J. Chem. Eng. Process Technol.* 5: 186.

Yang, K., Peng, J., Srinivasakannan, C., Zhang, L., Xia, H., Duan, X. 2010. Preparation of high surface area activated carbon from coconut shells using microwave heating. *Bioresource Technology*. 101(15), 6163-6169.

Yang, Y., Liu, J., Li, X., Liu, E., Li, Z., Qiu, M. 2013. New terpenoids from the roots of *Jatropha curcas*. *Chinese Science Bulletin*. 58(10): 1115-1119.

Yao, L., Han, C., Chen, G., Song, X., Chang, Y., Zang, W. 2012. A new asymmetric diamide from the seed cake of *Jatropha curcas* L. *Fitoterapia*. 83(8): 1318-1321.

Ye, J., Geng, Y., Zhang, B., Mao, H., Qu, J., Chua, N. H. 2014. The Jatropha FT ortholog is a systemic signal regulating growth and flowering time. *Biotechnology for Biofuels*. 7(1): 91.

Ye, J., Qu, J., Mao, H. Z., Ma, Z. G., Rahman, N. E. B., Bai, C., Chua, N. H. 2014. Engineering geminivirus resistance in Jatropha curcus. *Biotechnology for Biofuels*. 7(1): 149.

Zhang, X. Q., Li, F., Zhao, Z.-G., Liu, X. L., Tang, Y. X., Wang, M. K. 2012. Diterpenoids from the root bark of *Jatropha curcas* and their cytotoxic activities. *Phytochemistry Letters*. 5(4): 721-724.

Zhang, X. P., Zhang, M. L., Su, X. H., Huo, C. H., Gu, Y. C., Shi, Q. W. 2009. Chemical constituents of the plants from genus Jatropha. *Chemistry and Biodiversity*. 6(12): 2166-2183.

Zhao, Q., Wang, W., Wang, Y., Xu, Y., Chen, F. 2012. The effect of curcin from *Jatropha curcas* on apoptosis of mouse sarcoma-180 cells. *Fitoterapia*. 83(5): 849-852.

In: Jatropha Curcas
Editor: Gregory Medina

ISBN: 978-1-63483-089-8

Chapter 4

A TECHNOLOGY AND ENTREPRENEURSHIP INTERFACE MODEL FOR PROMOTING MULTIPLE USES OF JATROPHA FOR PRO-POOR DEVELOPMENT

Raphael M. Jingura [*] ***and Reckson Kamusoko***
Chinhoyi University of Technology, School of Agricultural Sciences and Technology, Chinhoyi, Zimbabwe

ABSTRACT

Jatropha curcas L. is a popular multipurpose crop that is extensively grown in many tropical countries. Its popularity has been ascribed to its acclaimed attributes as a non-food crop that grows well in marginal land not suitable for crop production, relatively low requirements of water, fertilizer and maintenance. It is arguably true that Jatropha can provide additional income in areas where there is plenty of land and labour is readily available. As such, Jatropha is now a component of farming systems in many developing countries. Traditionally, farming systems in developing countries have focused on the twin purposes of food and fibre production. The entrance of Jatropha adds a third dimension to this model. This arises mainly from the multiple uses of Jatropha, particularly energy production. The possibilities of exploitation of Jatropha for various uses have been explored and continue to be explored. This has largely been underpinned by technological innovations for processing different parts of Jatropha into value-added products. The major parts of Jatropha that can be exploited for multiple uses include the woody parts, seed, shell, husks, kernel, oil and press-cake. Press-cake is a by-product of oil extraction. Mature technologies exist for converting Jatropha components into useful products. The major focus has been on developing appropriate technologies for converting Jatropha components into first and second generation energy careers. The gamut of technologies available for conversion purposes ranges from biological to chemical methods. Uses in soap and fertiliser production, as food, medicine and pesticide have been explored. Interfacing the gamut of technologies available and the multiplicity of uses of Jatropha, provides an opportunity for promoting what can be called the Jatropha entrepreneurship system. There is no doubt that Jatropha can be more profitable promoting small to large-scale entrepreneurship if production is coupled with appropriate

[*] Corresponding Author: Email: rjingura@gmail.com.

management practices and seed oil is processed into other valuable products other than biofuel. Entrepreneurship is a major driver of economic development for large populations in developing countries. As such, crops like Jatropha provide opportunities through the various products that can be obtained from them. In this chapter we develop a model for promoting entrepreneurship based on the multiple uses of Jatropha and use of available technologies, emphasising the benefits to rural economies.

INTRODUCTION

Contemporary concerns about climate change, biodiversity and sustainability have led to numerous initiatives to explore new ways of exploiting natural resources for production of goods and services. Production ethos is now underpinned by concepts such as renewability, biodegradability and environment-friendliness of production systems and their products. The burgeoning world population, fast growing industries and energy-hungry new technologies compound the need to seek alternatives to existing limitations (Wani and Chander, 2012). Global initiatives in pursuit of alternative production systems for food, fibre and energy abound in the 21st century. It is in this realm that *Jatropha curcas* L. has been promoted as a suitable crop for sustainable production of multiple products that include energy.

Jatropha is a perennial oil-bearing shrub or small tree which originates from Mexico and Central America (Jubera et al. 2009). It is today well known in wild or semi-cultivated areas in many tropical and sub-tropical countries. As the Jatropha industry is still undergoing development, there is little consolidated knowledge on the acreage of Jatropha currently planted out. However, the multiplicity of the uses of Jatropha has been well elucidated. Literature abounds with knowledge on the technologies available for the conversion of different parts of Jatropha into various useful products. The Jatropha business is usually renowned for being a 'zero waste' system. This is ascribed to the usefulness of almost all its parts, from the structural parts to its seed oil. Jatropha has become an example for the tremendous hope placed in novel crops that "offer all the benefits of biofuels without the pitfalls" (Renner, 2007). It is known to deliver oilseeds from marginal lands under unfavourable soils and climates in (semi-) arid regions without compromising food production, diminishing natural resources or ecosystem services, such as carbon stocks and soil fertility (Wahl et al. 2012).

Jatropha, like most biomass, can be converted into useful products using different state-of- the- art processes. The gamut of available technologies, most of which are mature technologies, covers chemical, biological, thermal, mechanical or a combination of processes (McKendry, 2002). While some of these technologies are commercially viable, others are in the pilot phase or under research and development. The most dominant, on a commercial business scale, use of Jatropha has been the production of biodiesel (Chakrabarti and Prasad, 2012). Some of the other commercial uses of Jatropha include soap making, use as a bio-fertiliser and for medicinal and pesticidal applications.

The point worth noting is that where technology and enterprise interface, there is value creation. It is value creation, enabled by technology and other determinants that drive entrepreneurship. Although the Jatropha industry is still in its infancy, a typology of business models for the crop is already taking shape. A project's business model depends on its domestic and international target markets, that is, its products and target customers as well as its value-chain configuration (Wahl et al. 2012). There are several case studies on the business and economic aspects of Jatropha production. Good examples are cited in literature for India, Kenya, Mali and Tanzania (Borman et al. 2013; FAO, 2010; Romijn and Caniels, 2011).

Interfacing the gamut of technologies available and the multiple uses of Jatropha provides an opportunity for what can be called the Jatropha entrepreneurship system. There is no doubt that Jatropha can promote small to large-scale entrepreneurship if production is coupled with appropriate management practices and seed oil is processed into other valuable products other than biofuel. Entrepreneurship is a major driver of economic development for large populations in developing countries. As such, crops like Jatropha provide opportunities through the various products that can be obtained from them. In this chapter we develop a model for promoting entrepreneurship based on the multiple uses of Jatropha and use of available technologies, emphasising the benefits to rural economies.

COMPONENTS OF JATROPHA OF ECONOMIC IMPORTANCE

Jatropha is generally a small tree which can reach a height of 5 m, but can grow up to 10 m under favourable conditions (Kumar and Sharma, 2008). Parts of the plant of economic importance include roots, leaves, stems, bark and the fruit. The fruit can be disaggregated into its constituent parts which include the seed and shell. Figure 1 is an illustration of the different plant-parts of Jatropha.

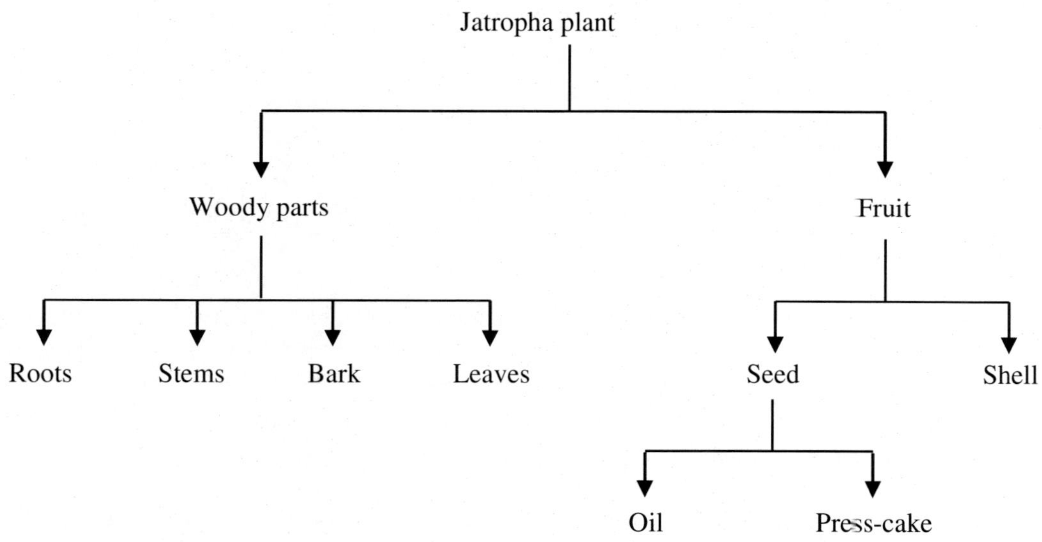

Figure 1. Different plant-parts of Jatropha.

Table 1. Chemicals isolated from different parts of Jatropha⁺

Various parts	Chemical composition	Reference
Stem bark	Saponins, steroids, tannins, glycosides, alkaloids and flavonoids β-amyrin, β-sitosterol and taraxerol	Igbinosa et al. (2009) Mitra et al. (1970)
Leaves	Flavonoids, apigenin, vitexin, isovitexin, sterol, stigmasterol, β-D-sitosterol, β-D-glucoside, sapogenins, alkaloids, triterpenae alcohol and 1-triacontal	Chhabra et al. (1990)
	A dimer of a tripene alcohol	Neuwinger (1994)
	A complex of 5-hydroxypyrrolidin-2-one and pyrimidine-2,4-dione	Staubmann et al. (1999)
	Stigmast-5-en-3β, 7β-diol, Stigmast-5-en-3β, 7α-diol, cholest-5-en-3β, 7β-diol, cholest-5-en-3fl, campesterol and 7-keto-β-sitosterol	Mitra et al. (1970), Khafagy et al. (1977), Hufford and Oguntimein (1987)
Latex	Tanins, saponins, wax and resin	Perry et al. (1980), Watt et al. (1962)
	Curcacycline A (a cyclic octapeptide)	Van den Berg et al. (1995)
	Curcacycline B (a cyclic nonapeptide)	Catherine et al. (1997)
	Curcin (a protease)	Nath and Dutta (1994)
Seeds	Curcin (a lectin)	Stirpe et al. (1976)
	Phorbol esters	Adolf et al. (1984), Makkar et al. (1997)
	Esterases (JEA and JEB) and lipase (JL)	Staubmann et al. (1999)
Roots	β-Sitosterol and its β-d-glucoside, marmesin, propacin, the curculathyranes A and B and the curcusones A–D. diterpenoids jatrophol and jatropholone A and B, the coumarin tomentin, the coumarino-lignan jatrophin as well as taraxerol	Naengchomnong et al. (1986, 1994)
	Steroids, alkaloids, saponins	Aiyelaagbe et al. (2007)
	Jatropholone A and B (diterpenoids), Jatropholol (diterpenoid)	Chen et al. (1988)

⁺Source: Prasad et al., 2012.

There is a large variability in different accessions of Jatropha from diverse agro-climatic regions (Kaushik et al. 2007). This variability manifests itself mainly in terms of chemical composition of the various plant parts. Jatropha produces fruits of various sizes with an average length and a diameter of about 2.5-3 cm and 2-3 cm, respectively (Heller, 1996). Each fruit contains 2-3 seeds that are black in colour, 2 cm long and 1 cm thick (Heller, 1996).

The fruit is made up of the shell (37.5%) and seeds (62.5%) (Abreu, 2008). The seeds are the most interesting parts of the plant from a potential commercial vantage point (Ahmed and Salimon, 2009). The Jatropha seed contains 30 to 40 percent oil which can be extracted by heat, solvents or by pressure. Oil extraction produces press-cake as a by-product.

Several chemicals can be extracted from different parts of Jatropha. The chemicals are shown in Table 1. These chemicals can be used in various industrial and non-industrial applications.

Economic Significance of Jatropha

A number of technologies are in vogue for conversion of different components of Jatropha into useful products.

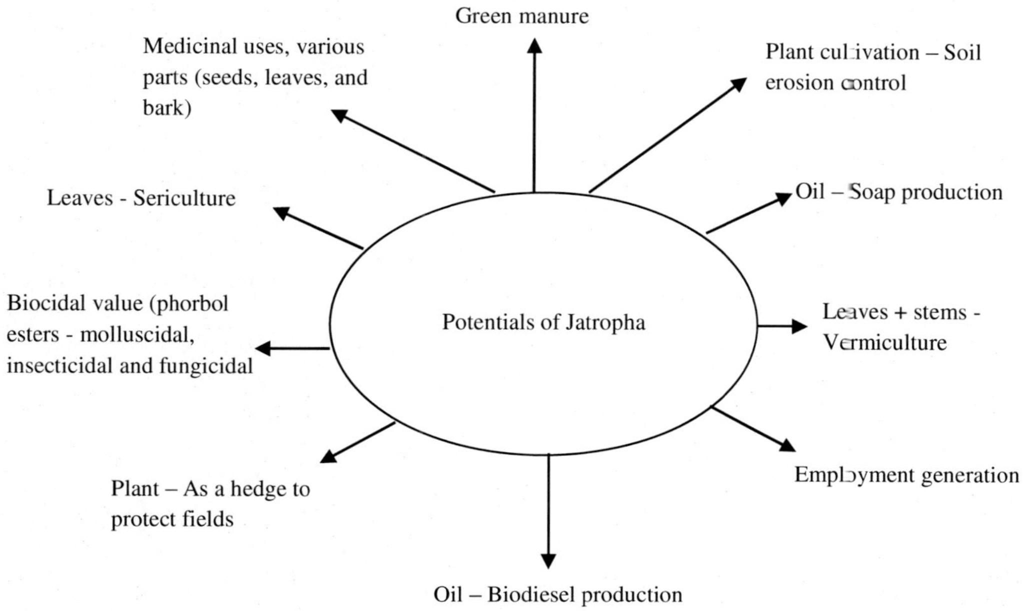

Figure 2. Economic significance of Jatropha (Kumar and Sharma, 2008).

All the different parts of Jatropha whether fresh or as a decoction have chemical properties useful for various purposes. Kumar and Sharma (2008) provided a good framework outlining the economic importance of Jatropha. This framework is shown in Figure 2. The most dominant and mainstream value of Jatropha is the production of biodiesel from seed oil. However, Jatropha has other multiple uses. These uses include use for medicinal and veterinary purposes (Akinpelu, 2009). It is also worth noting that Jatropha contains toxic compounds which include phorbol esters. These limit the use of Jatropha products for food purposes.

As a Source of Fuelwood and Charcoal

The components of Jatropha from which fuelwood can be derived are the woody materials in stems and fruit shells. Charcoal can also be produced from the woody material. Each Jatropha plant produces about 200 kg of biomass with a dry matter content of about 25% at about 7 years of growth (Benge, 2006). This yields a dry matter content of wood of about 80 t ha^{-1} in cuttings done once every 10 years. With an energy content of 15.5 MJ kg^{-1}, this has the potential to supply 1.2 PJ of energy (Jingura et al. 2010).

A mature technology for converting fuelwood into energy is combustion, commonly known as burning. Combustion is an ordinary way of using biomass as an energy source. It is a low cost, ease to handle and high reliability technology (European Commission, 2005). Combustion is a chemical reaction between fuel and oxygen and usually occurs in presence of air (Naik et al. 2010). The technology uses several process tools that range from simple systems such as domestic stoves, boilers, furnaces, steam turbines and turbo-generators to sophisticated systems like fluidised-bed reactors. The process equipment converts chemical energy stored in solid fuel biomass into heat, mechanical power or electricity. Carbon dioxide and water are the by-products of the exothermic reactions of combustion. Fuelwood combustion is the most common source of energy production in many poor rural households.

The carbonisation process produces charcoal. Carbonisation itself is a relatively low cost procedure and its application is common in some poor rural communities (FAO, 1985). Some people have suggested the use of press-cake for charcoal making. Other views are that press-cake is much more valuable to use as a fertiliser to ameliorate the impoverished soils with organic matter and nutrient content in order to increase crop production (Benge, 2006). There is an abundance of fragile soils and wasteland in areas where poor people reside in developing countries. Such areas can benefit from use of press-cake as a bio-fertiliser.

Briquetting is a common technology for producing high density energy carriers. Densification of shell material can provide shell briquettes which can be used for combustion to provide energy. Jatropha shell briquettes have been produced and used for combustion. With a calorific value of 11.1 MJ kg^{-1} and a yield of 1 t ha^{-1}, shells can supply 11.1 GJ of energy ha^{-1} (Sotolongo et al. 2007).

Use of Jatropha as fuelwood and charcoal is limited by two main constraints. The first is that Jatropha wood is light with a density of 0.35 and is not good for both fuelwood and charcoal (Benge, 2006; Islam et al. 2011). The wood burns too quickly. Therefore, the use of Jatropha as fuelwood or charcoal can only be minimal (Sotolongo et al. 2007). However, the islands of Cape Verde have used Jatropha as the only reliable source of fuelwood due to unavailability of suitable species (Brittaine and Lutaladio, 2010). The second issue is converting Jatropha seed shells and wood into fuelwood, charcoal or briquettes would be economically feasible only if there are large sources of materials from Jatropha plantations. Unless the size of Jatropha plantations escalates to large proportions, the general conclusion seems to be that Jatropha is not much of value as both fuelwood and charcoal (Benge, 2006).

As a Source of Oil

Seed oil is the component of Jatropha with the highest economic importance. Jatropha seed oil has multiple uses which include the following:

- Raw oil for diesel engines, lighting and cooking purposes
- Soap making
- Production of biodiesel

Raw Oil for Diesel Engines, Lighting and Cooking Purposes

Raw oil has been used as a substitute for petro-diesel both in modified and unmodified diesel engines. The oil from Jatropha is a potential fuel substitute as shown by its use in many applications.

Table 2. Selected fuel properties of raw Jatropha oil

Properties	Values	Reference
Calorific value (MJ kg^{-1})	39.07	Abreu (2008)
Kinematic viscosity at 40°C (mm^2 sec^{-1})	41.51	Kywe and Oo (2009)
Flash point (°C)	229.3	Shambhu et al. 2013
Pour point (°C)	6	Dubey et al. (2011)
Sulphur content (% wt)	0.0002	Chalatlon et al. (2011)
Specific gravity	0.87	Kywe and Oo. (2009)
Acid value mg/NaOH/g	2.16	Dubey et al. (2011)
Ash content (%)	0.01	Shambhu et al. (2013)

The fuel properties of Jatropha oil are shown in Table 2. The use of raw oil as a fuel has not shown satisfactory results due to its high viscosity (Abreu, 2008; Shahid and Jamal, 2008). The viscosity of raw oil is very high (41.5 mm^2 sec^{-1}) compared to the ASTM standard requirement for diesel fuel which is 1.3-4.1 mm^2 sec^{-1} (Abreu, 2008). This high viscosity of the raw oil causes problems in its use in diesel engines. These include reducing the fuel atomisation and increasing fuel spray, which would be responsible for engine deposits, injector coking, piston ring sticking and thickening of lubricating oil (Abreu, 2008; Shahid and Jamal, 2008). nDespite the problems caused by its chemical properties for direct use in diesel engines, raw oil has some other uses. Jatropha oil can be used in diesel engines directly or by blending it with methanol (Gubitz et al. 1999). Raw oil has been used in slow speed stationary diesel engines with success (Tomomatsu and Brent, 2007). An example is the Lister-type of engines. However, these will need to be modified to account for high oil viscosity and low absorbance capacity (Tomomatsu and Brent, 2007). This type of engine has wide use in many scenarios in the rural and poor communities as shall be explored later in this chapter. Jatropha oil can be used as an illuminant because it burns slowly with little or without emitting smoke. Some reports suggest that Jatropha nuts can be strung on grass and be burned like candlenuts, and the oil can be used to make candles (Benge, 2006).

However, Jatropha oil cannot fully substitute fuels like kerosene for use in cooking and lighting. It cannot be used directly in conventional kerosene stoves or lamps (Benge, 2006). High ignition temperatures and viscosity as compared to kerosene (50-55 ^{0}C, and 2.2 10^{-6} m^2/s, respectively) mean that Jatropha oil will not burn as well, and would clog up all the tubes and nozzles in a conventional stove or lamp (Benge, 2006). This point is very important considering that one of the challenges in rural and poor households is provision of affordable energy carriers for space heating, lighting and cooking. The limitations of Jatropha oil as a substitute of fossil fuels can be overcome by technological interventions. For example, degumming and transesterification processes reduce the viscosity by 19.8% and 74.8%, respectively (Dubey et al. 2011). According to Dubey et al. (2011) heating of degummed oil at 70 ^{0}C reduced the viscosity by 6.8 times when compared with viscosity at 10 ^{0}C. Their work showed that the degummed oil at 70 ^{0}C becomes viscous at par with fossil diesel, hence could be used directly in IC engines. Transesterification shall be discussed later in this chapter. The bottom line is that technological interventions make Jatropha oil usable as a petro-diesel substitute.

Soap Making

Extensive soap production from raw Jatropha oil has been ascribed to a very high saponification value (202.34 mg of KOH/g oil) of the oil (Mohammed-Dabo and Ahmad, 2013). Soap can be produced from raw Jatropha oil or from glycerine produced as a by-product of biodiesel production. In either case the process produces a soft, durable soap. The process is a simple one, well adapted to household or small-scale industrial activity (Kumar and Sharma, 2008). Research carried out in India has shown that with a mixture of 75% hydrogenated Jatropha oil, 15% refined and bleached Jatropha oil, and 10% coconut oil, a soap can be produced with lathering values equivalent to regular toilet soap (Benge, 2006). Using an oil content of 30%, pressing of 10 kg of Jatropha seed produces 3 kg of oil which can be converted to soap. Soap making is a common activity in village or cottage industries in developing countries. It requires minimum investment with good returns (Benge, 2006).

Biodiesel from Jatropha

Biodiesel is produced from oils through a process called transesterification. Biodiesel is the term widely used for alkyl esters of fatty acids made from vegetable oils or animal fats. The most dominant economic use of Jatropha has been the production of biodiesel from its seed oil. For reasons stated above, raw Jatropha seed oil is not a suitable substitute of petro-diesel unless it is processed to reduce its high viscosity. Biodiesel requires very little or no engine modifications up to 20% blend and minor modification for higher percentage blends (Kumar and Sharma, 2008). Table 3 shows some of the performance characteristics of Jatropha biodiesel.

Use of Press-Cake

As stated earlier, press-cake is a by-product of oil extraction by mechanical expelling. About 70-75% (Singh et al. 2008) of the original seed weight, primarily in the form of carbohydrates and proteins (Brittaine and Lutaladio, 2010) remains as press-cake after oil extraction. Press-cake contains a substantial amount of oil depending on the extraction process. This oil influences the gross energy productivity of the press-cake (Achten et al. 2008). In large-scale operations, disposal of large amounts of press-cake can be a problem since it cannot be used as an animal feed in its unprocessed form. This is due to presence of toxic constituents like phorbol esters, curcin and the presence of antinutrients and toxins (Martínez-Herrera et al. 2006). Jatropha produces approximately 1 t ha^{-1} of press-cake after oil extraction (Pandey et al. 2012). Alternate uses of press-cake are derived from its chemical composition. Proximate composition analysis showed that it has 94% total solids out of which 93% is volatile solids.

Table 3. Performance characteristics of Jatropha biodiesel

Variable	Characteristic	Reference
Viscosity	Exhibits similar kinematic viscosity to petro-diesel	Demirbas (1998); Marchetti et al. (2007)
Biodegradability	Excellent biodegradability	Speidel et al. (2009)
Sulphur and aromatic compounds (emission)	Does not contain any sulphur or aromatic compounds, thus lower level of toxic emissions	Benge (2006); USEPA Report (2002)
NO_x emissions	A little higher compared to petro-diesel	Chakrabati and Prasad (2012)
Other emissions	Produces less unburnt hydrocarbons, particulate matter carbon monoxide than petro-diesel	Chakrabati and Prasad (2012)
Some properties	Lower calorific value, lower volatility and slightly corrosive against brass and copper than petro-diesel	Chakrabati and Prasad (2012)
Engine performance	Shows satisfactory engine performance	Kumar and Sharma (2008)

The cake is also high in organic matter and has the potential to be utilised as a feedstock for biogas production (Singh et al. 2008). Use of biogas slurry as a fertilizer is a possibility that can be explored and some work is being done.

The solid-state fermentation of Jatropha press-cake has shown that it could be a good source of low cost production of industrial enzymes such as protease and lipase (Mahanta et al. 2008; Pandey et al. 2012). Press-cake may also be converted to briquettes for domestic or industrial combustion (Pandey et al. 2012).

Production of biogas by anaerobic digestion of Jatropha press-cake has been demonstrated (Francis et al., 2005). Bio-methanation is a mature energy technology that can be used to convert press-cake into biogas. Jatropha press-cake is a good substrate for biogas production (Foidl, 1996). Biogas is a robust fuel that can be used to supply heat, electricity, process steam and methanol.

The press-cake contains adequate amounts of carbohydrates and proteins for digestion by anaerobic bacteria to produce large amounts of biogas. Methane production levels of 0.5 to 0.6 m^3 kg^{-1} dry matter can be obtained from anaerobic digestion of Jatropha press-cake as a single substrate (Visser and Adriaans, 2007). It has also been observed that more than 60 days residence time is required to convert the digestible fraction completely into biogas (Visser and Adriaans, 2007).

Table 4. Nutritional analysis of oilseed cakes and manure (%) (Delgado and Parado, 1989)

Property	*J. curcas* oil cake	Neem oil cake	Cow manure
Nitrogen	3.2–4.44	5.0	0.97
Phosphorus	1.4–2.09	1.0	0.69
Potassium	1.2–1.68	1.5	1.66

One of the most valuable uses of press-cake is production of fertiliser. Press-cake is generally described as an excellent fertiliser material. Press-cake contains in the range of 50-62% protein and thus is an excellent source of N. The nutrient content of Jatropha press-cake fertiliser is comparable to most other organic fertilisers. This is shown in Table 4.

Experiments with wheat in India have shown that substitution of inorganic fertiliser by *Jatropha* press-cake significantly improved yields (Ghosh et al. 2012). In this study, substitution rates of 50, 75 and 100% of inorganic fertiliser with Jatropha press-cake produced higher yields over the control (100% recommended rate of inorganic fertiliser). There are many other experiments that have shown the value of Jatropha press-cake as a fertiliser. A logical conclusion is that there is enormous scope for use of Jatropha press-cake as an organic fertiliser. A net effect is possible reduction in the life cycle greenhouse gas emissions of agricultural crops grown under use of Jatropha press-cake fertiliser.

Use as Pesticide

Crop pests pose a challenge to crop production in all regions of the world. On the ledger of crop production systems, crop protection is a major cost driver. Empirical evidence shows that crop protection costs have increased over the years (Kumar and Sharma, 2008). Because of increasing problems associated with the use of toxic synthetic insecticides, there is need for development of safer alternative crop protectants (Rao et al. 2012). Jatropha oil and aqueous extract from different parts have potential as pesticides. All parts of Jatropha show insecticidal properties against pests in many crops. Extracts from different parts of Jatropha contain various compounds as shown in Table 1. Phorbol esters have been found to be the major toxic compounds in Jatropha. Pesticidal activities of Jatropha oil containing phorbol esters have been reported in *Manduca sexta, Helicoverpa armigera, Aphis gossyii, Pectinophora gossypielaa, Empoasca biguttula* and others (Wink et al. 1997). The oil and its aqueous extract has been shown to be effective against cotton insect pests including cotton bollworm as well as pests of pulse, potato and corn (Kumar and Sharma, 2008).

Medicinal Uses

Jatropha is widely known as a source of different medicinals for treatment of a variety of ailments (Benge, 2006). In fact, Jatropha is known as the purging nut because it has purgative/laxative effects. All parts of Jatropha are reported to have medicinal properties as shown in Table 5.

Other Potential Uses

Records of the prospective uses of Jatropha are inexhaustible. Jatropha is listed as a honey plant. Its flowers can attract bees, therefore it is more likely to provide a substantial income generation to the poor from honey production. The bark of Jatropha exudes a dark blue dye that has the credentials to support the clothing and fishing industry. The extract can be used for colouring clothes, fishing nets and lines (Islam et al. 2011). Great opportunities for further research in this area are obtainable. Jatropha leaves can be consumed by silkworm as part of its diet. As such, silk is produced from silkworm and can be woven into textile.

The supply of safe water that is free from toxic pollutants has been problematic in the developing world. Several available technologies to remove these contaminants from water have challenges of high capital and operational demands. These problems can be overcome through adsorption by the use of Jatropha seed husk activated carbon (Pandey et al. 2012). Jatropha oil can be used to soften leather and lubricate small equipment such as chain saws (Abdul Khalil et al. 2013).

Jatropha-Based Entrepreneurship

The multiple uses of Jatropha create vast opportunities for enterprise development. Various technologies are available to convert Jatropha components into multiple products. Jatropha has been widely promoted as a crop for pro-poor development. While the aim of pro-poor development is to increase economic benefits to the poorer members of society, such development should not unduly threaten food or water security, reduce access to land or create poor working conditions (Brittaine and Lutaladio, 2010). The thesis promoted in this chapter is that value addition is the most effective way to derive maximum benefits from Jatropha at the household level in poor communities. Small farmers need not just produce and sell Jatropha seed for mainstream value addition to which they are not part of.

Cognisant of the simplicity of most of the technologies available for conversion of Jatropha components into value added products, it is possible to promote value addition at both household and small enterprise levels in poor communities. It is this interface of technology adoption and multiple uses of Jatropha that creates opportunities for enterprise development. In view of this, the model depicted in Figure 2 on the economic importance of Jatropha can be modified to include intervening technologies for value addition. A variant of this model is shown in Figure 3.

Soap Making Entrepreneurship

As stated earlier in this chapter, soap making is a simple and affordable process. Benge (2006) actually described the process as village or cottage industry technology. This chapter will not provide a financial analysis of the soap-making business but only consider the technical feasibilities. The soap is made by adding a solution of sodium hydroxide (caustic soda) to Jatropha oil. This simple technology has turned soap making into a viable small-scale rural enterprise appropriate to many rural areas of developing countries (Brittaine and Lutaladio, 2010).

Table 5. Uses of different parts of Jatropha in medicines

Plant part	Therapeutic Uses	Reference
Whole	Wounds, allergies, burns, cuts	Heller (1996), Kaushik and Sharma (2004)
	Dysphonia, dyscrasia	Kirtikar and Basu (1980)
	Diarrhoea	Patil (2005)
	Antibiotic, insecticidal, toothache	Balee (1994)
Leaf	Asthma, bronchitis, as an analgesic, scorpion sting, emmenagogue	Nayak and Patel (2009)
	Purgative	Mongkolvisut et al. (2006)
	Constipation, vertigo, diarrhoea, skin diseases, mouth blister, cancer	Burkill (1994)
	Stomachache, eczema, carbuncles, itches, swelling, venereal disease, blood purifier	Banerji et al. (1993)
Exudates	Fever, borne fracture	Comerford (1996)
	Mouth blister	Flores and Ricalde (1996)
	Anti-inflammatory	Nayak and Patel (2009)
Sap/ latex	Toothache, stypic	Heller (1996), Kaushik and Sharma (2004)
	Wounds	Kirtikar and Basu (1980), Burkill (1994)
	Chest pain, retching, vomiting, stomach ache, eye infections, purgative, haemostatic	Miller and Morris (2004)
Fruit and seed	Abdominal complains, dysentery, urinary discharge, fistula, heart disease, antihelmintic	Kirtikar and Basu (1980)
Seed oil	Purgative	Asprey and Thornton (2005)
	Rheumatism, skin diseases, eczema	Heller (1996)
	Anti-ulcer, leprosy, ringworm	Banerji et al. (1993)
	Hair delouse	Schmelzer and Gurib-Fakim (2008)
	Arbortifacient	Kirtikar and Basu (1980
Root	Microbial infections Anti-ulcer Gonorrhoea, urinary discharge Invigorating drink	Aiyelaagbe (2000) Kirtikar and Basu (1980) Burkill (1994) Schmelzer and Gurib-Fakim (2008)
	Leprosy, antidote	Kirtikar and Basu (1980)
Stem/bark	HIV, tumour	Heller (1996), Kaushik and Sharma (2004)

Source: Sabandar et al., 2013.

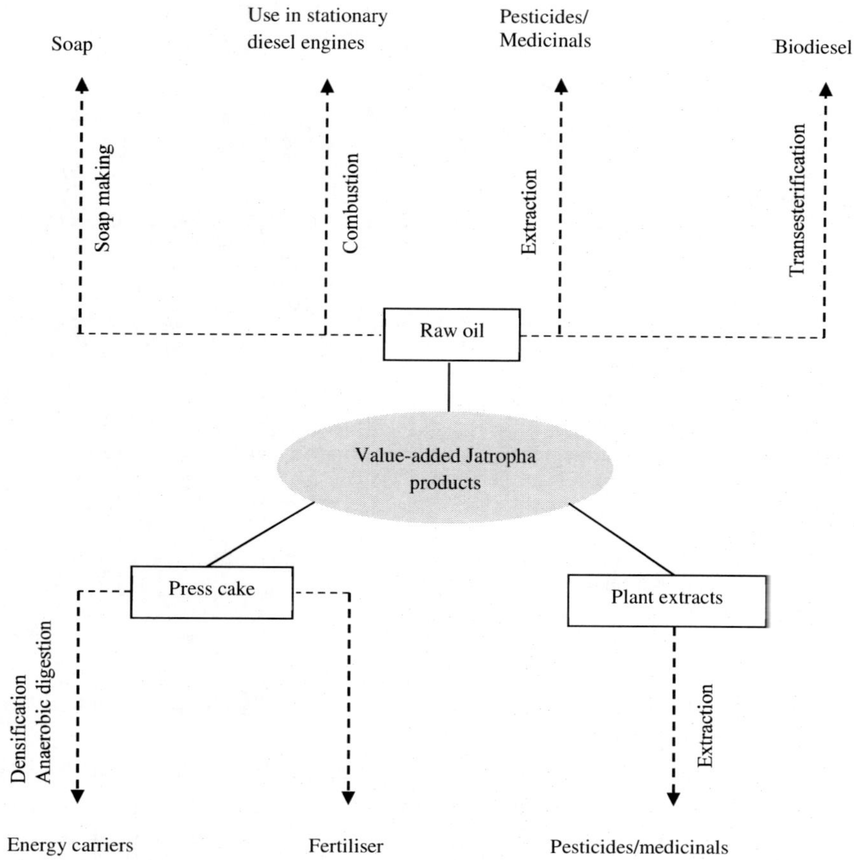

Figure 3. Value-added Jatropha products.

The attraction of Jatropha soap is that it has medicinal properties and can be priced higher than ordinary soap. The dynamics of production show that 4.7 kg of soap can be produced from 13 litres of Jatropha oil (Henning, 2004). There have been concerns raised with the quality of soap produced in small businesses.

These technical challenges can be overcome by appropriate training and modernisation of production methods. A good substantive story of soap making from Jatropha oil comes from Tanzania and is shown in Box 1.

Biofuel Entrepreneurship

Energy poverty is rampant in many rural communities in developing countries. Biofuels offer an opportunity to reduce energy poverty in rural and poor communities in developing countries.

The cultivation of Jatropha for seed production expands livelihood options with the opportunity to earn income for smallholder growers, oil mill out-growers and members of community plantation schemes or through employment on private enterprise Jatropha plantations (Brittaine and Lutaladio, 2010).

Box 1

Twenty-one-year-old is the founder of Jatropha Soap Production in Tanzania, which produces handmade soap with medicinal uses

The idea of producing *Jatropha* soap came to her when she and her brother attended an agricultural festival in 2012 and were introduced to some of the medicinal and cosmetic uses of the Jatropha plant in her country. For example, the plant extract has healing properties for a number of skin problems, such as ringworm and dandruff.

"The plant has seeds and we extract oil from the seeds and we use the oil to make soaps. We have *Jatropha* plants in our country but people are not using it," explained Silayo. "So I thought that was an opportunity and started making the soap."

After researching how to make soap and raising finance from family to buy oil and equipment, Silayo and her brother went into producing *Jatropha* soap from a room they have rented. Today they produce about 1,000 bars of soap a month and have one other full-time employee.

Source: Kate Douglas – 15 January 2014
http://www.howwemadeitinafrica.com/21-year-old-tanzanian-talks-about-her-soap-manufacturing-business/34147/

In addition to this, it is value addition in terms of producing energy carriers that enhances income from Jatropha production.

Opportunities for value addition in Jatropha-based entrepreneurship in terms of producing biofuels are shown in Figure 3. Relatively simple technologies are used to produce fuels such as biodiesel, biogas and raw oil. Demand for energy in rural communities is for cooking, lighting, powering engines for milling, irrigation and power-cutting. These needs create a market for cheaper biofuels in these areas. Both Jatropha oil and press-cake produce good fuels. Table 6 provides a business case for Jatropha-derived energy carriers.

Information in Table 6 buttresses the opportunities for addressing energy poverty in poor rural communities by producing Jatropha and adding value to its products. What is needed is to inculcate a business culture in rural livelihoods in order to exploit these opportunities. There is an opportunity to increase the value of the natural resource asset base of the rural poor by utilizing Jatropha's ability to grow on poor and saline soils in dry regions (Brittaine and Lutaladio, 2010).

For this to be realised, issues to do with economic viability of Jatropha production need to be addressed by continuous improvement in the agronomy of the crop.

Organic Fertiliser Entrepreneurship

Farming systems in rural areas in developing countries are dominated by crop production. Production of crops consumes copious amounts fertilisers, particularly inorganic fertilisers. Inorganic fertilisers are major drivers of greenhouse gas emissions in life cycle analysis of crops and their products. Considering production of crop-based industrial feedstock, the main driver of greenhouse gas emissions is fertiliser production and use. Thus, organic fertilisers offer an opportunity to mitigate effects of climate change. This is a good thing.

About 70% of Jatropha seed remains as press-cake after oil extraction. Conversion of this press-cake into an organic fertiliser has been reported to produce good results with various crops. Given that one of the limiting factors in crop production in rural communities is inadequate use of fertilisers due to inhibiting costs, press-cake fertiliser is a good supplement. Box 2 encapsulates the commercial sense of Jatropha press-cake fertiliser.

Table 6. Technical feasibility and use of biofuels in poor rural communities

Fuel	Production technology	Technical feasibility+	Market legitimacy
Raw oil	Extraction	***	– Can substitute petro-diesel in stationary engines – Can be used to run diesel engines for grinding, power cutting and irrigation – Reduces tedious work in grinding for women
Biodiesel	Transesterification	*	– Substitutes diesel in all diesel engines – Produces glycerine which has other uses
Biogas (press-cake)	Anaerobic digestion	***	– Good cooking and lighting fuel – Reduces energy poverty for domestic purposes – Clean cooking in smoke-free environment
Briquettes (press-cake)	Densification	**	– Substitute solid fuel
+Feasibility to be done in rural communities (*** very feasible, **moderately feasible, *feasible)			

Box 2 *Value of Jatropha press-cake fertiliser*

- Rich in N, P and K which are expensive nutrients in inorganic fertilisers
- Organic fertiliser in high demand in organic agriculture
- Large market exists for organic fertilisers
- Copious amounts produced during oil extraction
- Has many benefits to the soil other than nutrient enrichment

CONCLUSION

Jatropha as a crop presents multifunctional properties and an array of utilities that can promote entrepreneurship in economically vulnerable populations. The basic model of Jatropha-based entrepreneurship in centred on the interface of a plethora of conversion technologies and multiplicity of uses of Jatropha.

Most of the conversion technologies in vogue are fairly feasible for application at household and small business levels. Thus, for pro-poor development, Jatropha production needs to go beyond production of seed for sale to mainstream players. The extension should be value addition to produce more profitable products such as energy carriers, bio-fertilisers, soap, medicines and pesticides.

The technologies for this to happen exist. What needs to be done is to promote their use in poor communities. To this end, this chapter has provided a model for promoting Jatropha-based entrepreneurship with the use of available technologies for pro-poor development.

References

Abdul Khalil, H.P.S., Sri Aprilia, N. A., Bhat, A. H., Jawaid, M., Paridah, M. T. & Rudi, D. (2013). "A. Jatropha biomass as renewable materials for biocomposites and its applications," *Renewable and Sustainable Energy Reviews*, 22, 667-685.

Abreu, F. (2008). Alternative By-products from Jatropha. In: International Consultation on Pro-poor Jatropha Development. 10–11 April 2008. Rome. IFAD. Available from: http://www.ifad.org/events/jatropha/.

Achten, W.M.J., Verchot, L., Franken, Y.J., Mathijs, E., Singh, V.P., Aerts, R., & Muys, B. (2008). Jatropha Bio-Diesel Production and Use. *Biomass and Bioenergy,* 32(12), 1063-1084.

Ahmed, W.A., & Salimon, J. (2009). Phorbol Ester as Toxic Constituents of Tropical *Jatropha Curcas* Seed Oil. *European Journal of Scientific Research,* 31(3), 429-436.

Akinpelu, D.A., Aiyegoro, O.A., & Okoh, A.I. (2009). The Bioactive Potentials of Two Medicinal Plants Commonly Used as Folklore Remedies among Some Tribes in West Africa. *African Journal of Biotechnology,* 8(8), 1660-1664.

Benge, M. (2006). Assessment of the potential of *Jatropha curcas* (biodiesel tree) for energy production and other uses in developing countries. Available from: http:// www. echotech.org/mambo/index.php?option=com_docman&task=doc_view&gid=179 & Itemid=68.

Borman, G., von Maltitz, G., Tiwari, S., & Scholes, M. (2013). Modelling the economic returns to labour for Jatropha cultivation in southern Africa and India at different local fuel prices. *Biomass and Bioenergy,* 59, 70-83.

Brittaine, R., & Lutaladio, N. (2010). Jatropha: A Smallholder Bioenergy Crop**.** The Potential for Pro-Poor Development. *Integrated Crop Management,* 8, 77-96.

Chakrabarti, P.P., & Prasad, R.B.N. (2012). Biodiesel production from *Jatropha curcas* oil. In N. Carels, M. Sujatha, &B. Bahadar (Eds.), Jatropha, Challenges for a New Energy Crop: Volume 1: Farming, Economics and Biofuel (pp. 463-490). New York: Springer Science + Business Media.

Chalatlon, V., Roy, M.M., Dutta, A., & Kumar, S. (2011). "Jatropha oil production and an experimental investigation of its use as an alternative fuel in a DI diesel engine". *Petroleum Technology and Alternative Fuels,* 2, 76–85.

Delgado, M.J.L., & Parado Tejeda, E. (1989). Potential multipurpose agro-forestry crops identified for the Mexican Tropics. In G.E. Wickens, N. Haq, & P. Day (Eds.), New Crops for Food and Industry (pp. 166–173). London: Chapman and Hall.

Dermibas, A. (1998). Fuel properties and calculation of higher heating values of vegetable oils. *Fuel,* 77, 1117-1120.

Dubey, A.K., Sarviya, R.M., & Rehman, A. (2011). Characterization of Processed Jatropha Oil for use as Engine Fuel. *Current World Environment,* 6(1), 101-107.

European Commission. (2005). Biomass: green energy for Europe. Luxembourg: Office for Official Publications of the European Communities, 1-49.

Foidl, N., Foidl, G., Sanchez, M., Mittelbach, M. & Hackel, S. (1996). *Jatropha curcas* L. as a source for the production of biofuel in Nicaragua, Bioresources *Technology,* 58, 77-82.

Food and Agricultural Organisation. (1985). Industrial Charcoal Making. FAO Forestry paper 3. FAO, Rome, Italy.

Food and Agricultural Organisation. (2010). Jatropha: a smallholder bioenergy crop. The Potential for pro-poor development. Rome, Italy: Food and Agriculture Organization of the United Nations (FAO).

Francis, G., Edinger, R., & Becker, K. (2005). A concept for simultaneous wasteland reclamation, fuel production, and socio-economic development in degraded areas in India: Need, potential and perspectives of *Jatropha* plantations. *Natural Resources Forum,* 29, 12-24.

Ghosh, A., Chikara, J., & Chaudhary, D.R. (2012). Value-addition of Jatropha cake and its utilisation as manure in Jatropha and other crops. In N. Carels, M. Sujatha, & B. Bahadar (Eds.), Jatropha, Challenges for a New Energy Crop: Volume 1: Farming, Economics and Biofuel (pp. 355- 368). New York: Springer Science + Business Media.

Gubitz, G.M., Mittelbech, M., & Trabi, M. (1999). Exploitation of tropical oil seed plant *Jatropha curcas* L. Bioresources. Technology, 67, 73–82.

Henning, R. K. (2004). *Jatropha curcas* L. in Africa. Case Study. Weissenberg, Germany, Bagani.

Heller, J. (1996). Physic nut. *Jatropha curcas.* L. Promoting the conservation and use underutilised and neglected crops. 1. Institute of Plant Genetics and Crop Plant Research. Gaterslebenl International Plant Genetic Resources Institute, Rome.

Igbogidi, O.M., Akparobi, S.O., & Eruotor, P.G. (2013). Health and environmental benefits of *Jatropha curcas* Linn. *Unique research Journal of Agricultural Sciences,* 1(5), 076-079.

Islam, A.K.M.A., Yaakob, Z., & Anuar, N. (2011). Jatropha: A multipurpose plant with considerable potential for the tropics. *Scientific Research and Essays,* 6(13), 2597-2605.

Jingura, R.M., Musademba, D., & Matengaifa, R. (2010). An evaluation of utility of *Jatropha curcas* L. as a source of multiple energy carriers. *International Journal of Engineering, Science and Technology,* 2(7), 115-122.

Jubera, M.A., Janagoudar, B., Biradar, D., Ravikumar, R., Koti, R., & Patil, S. (2009). Genetic diversity analysis of elite *Jatropha curcas* (l.) genotypes using randomly amplified polymorphic DNA markers. *Karnataka Journal of Agricultural Science,* 22, 293-295.

Kaushik, N., Kumar, K., Kumar, S., Kaushik, N., & Roy, S. (2007). Genetic variability and divergence studies in seed traits and oil content of Jatropha (*Jatropha curcas* L.) accessions. *Biomass and Bioenergy,* 31, 497-502.

Kumar, A., & Sharma, S. (2008). An evaluation of multipurpose oil seed crop for industrial uses (*Jatropha curcas* L.): A review. Industrial Crops and Products. Available from: http://www.highbeam.com/doc/1P3-1499175641.html:

Kywe, T.T. & Oo, M.M. (2009). Production of biodiesel from Jatropha oil (*Jatropha curcas*) in pilot plant. Proceedings of World Academy of Science, Engineering and Technology, 38, 481-487.

Mahanta, N., Gupta, A., & Khare, S.K. (2008). Production of protease and lipase by solvent tolerant Pseudomonas aeruginosa PseA in solid-state fermentation using *Jatropha curcas* seed cake as substrate. *Bioresources Technology,* 99, 1729–1735.

Marchetti, J.M., Miguel, V.U., & Errazu, A.F. (2007). Possible methods for biodiesel production. *Renewable and Sustainable Energy Reviews*, 11, 1300-1311.

Martínez-Herrera, J., Siddhuraju, P., Francis, G., Dàvila-Ortíz, G., & Becker, K. (2006). Chemical composition, toxic/antimetabolic constituents, and effects of different treatments on their levels in four provenances of *Jatropha curcas* L. From Mexico. *Food Chemistry,* 96, 80-89.

McKendry, P. (2002). Energy production from biomass (part 2): Conversion technologies. *Bioresource Technology,* 83, 47-54.

Mohammed-Dabo, I.A., & Ahmad, M.S. (2013). Single phase (cyclohexane-assisted) transesterification of *Jatropha curcas* seed oil using full factorial design of experiment. *International Journal of Renewable and Sustainable Energy,* 2(5), 191-197.

Naik, S.N., Goud, V.V., Rout, P.K., & Dalai, A.K. (2010). Production of first and second generation biofuels: A comprehensive review. *Renewable and sustainable energy reviews,* 14, 578-597.

Pandey, V.C., Singh, K., Singh, J.S., Kumar, A., Singh, B., & Singh, R.P. (2012). *Jatropha curcas*: A potential biofuel plant for sustainable environmental development. *Renewable and Sustainable Energy Reviews,* 16, 2870-2883.

Prasad, D.M.R., Izam, A., & Khan, M.R. (2012). *Jatropha curcas*: Plant of medical benefits. *Journal of Medicinal Plants Research,* 6(14), 2691-2699.

Rao, M.D., Anitha, S., Aravinda, A., Karunakar, B., & Devanna, N. (2012). Biopepticidal properties of seed, seed cake and oil of *Jatropha curcas* L. Against the polyphagous lepidopteran pest Helicoverpa armigera. In N. Carels, M. Sujatha, & B. Bahadar (Eds.), Jatropha, Challenges for a New Energy Crop: Volume 1: Farming, Economics and Biofuel (pp. 369-382). New York: Springer Science + Business Media.

Renner, R. (2007). Green gold in a shrub. Entrepreneurs target the Jatropha plant as the next big biofuel. *Scientific American,* 296(6), 20-23.

Romijn, H., & Caniëls, M. (2011). The Jatropha biofuels sector in Tanzania 2005–2009: evolution towards sustainability? *Research Policy,* 40(4), 618-636.

Sabander, C.W., Ahmat, N., Jaafar, F.M., & Sahidin, 1. (2013). Medicinal property, phytochemistry, pharmacology of several Jatropha species (Euphorbiaceae): A review. *Phytochemistry,* 85, 7-29.

Shahid, E.M. & Jamal, Y. (2008). A review of biodiesel as a vehicular fuel. *Renewable and Sustainable Energy Reviews,* 12(9), 2484-2494.

Shambhu, V.B., Bhattacharya, T.K., Nayak, L.K., & Das, S. (2013). Studies on characterisation of raw Jatropha oil and its biodiesels with relevance of diesel. *International Journal of Emerging Technology,* 3(4), 48-54.

Singh, R.N., Vyas, D.K., Srivastava, N.S.L., & Narra, M. (2008). SPRERI experience on holistic approach to utilise all parts of *Jatropha curcas* fruit for energy. *Renewable energy,* 33, 1868-1873.

Sotolongo, J.A., Beatón, P., Diaz, A., de Oca, S.M., del Valle, Y., Pavón, S.G., & Zanzi, R. (2007). *Jatropha curcas* L. as a source for the production of biodiesel. A Cuban experience. *In:* 15th European Biomass Conference and Exhibition", 7-10 May, Berlin, Germany, 2007, 2631-2633.

Speidel, H.K., Lightner, R.L,. & Ahmed, I. (2000). Biodegradability of new engineered fuels compared to conventional petroleum fuels and alternative fuels in current use. *Applied Biochemistry and Biotechnology,* 84-86,879-897.

Tomomatsu, Y., & Brent, S. (2007). *Jatropha curcas* biodiesel production in Africa: Economics and potential value chain development for smallholder farmers. WP54. Nairobi, World Agroforestry Centre.

USEPA Report. (2002). A comprehensive analysis of biodiesel impacts on exhaust emission. Draft technical report, USEPA.

Visser, J., & Adriaans, T. (2007). Anaerobic digestion of *Jatropha curcas* press cake. Fact Fuels Foundation. Eindhoven, Netherlands.

Wahl, N., Hildebrandt, T., & Moser, C., Lüdeke-Freund, F., Averdunk, K., Bailis, R., Barua, K., Burritt, R., Groeneveld, J., Klein, A., Kügemann, M., Walmsley, D., Schaltegger, S., & Zelt, T. (2012). Insights into Jatropha projects worldwide: Key facts and figures from a global survey. Lüneburg: Centre for Sustainability Management.

Wani, S. P., Osman, M., D'Siva, E., & Sreedevi, T.K. (2006). "Improved Livelihoods and Environmental Protection Through Biodiesel Plantations in Asia." *Asian Biotechnology and Development Review,* 8 (2), 11–29.

Wani, S.P., & Chander, G. (2012). *Jatropha curcas* Biodiesel, Challenges and Opportunities: Is it a panacea for energy crisis, ecosystem service and rural livelihoods? In N. Carels, M. Sujatha, & Bahadar, B. (Eds.), Jatropha, Challenges for a New Energy Crop: Volume 1: Farming, Economics and Biofuel pp. 311- 331. New York: Springer Science + Business Media.

Wink, M., Koschmieder, C., Sauerwein, M., & Sporer, F. (1997). Phorbol esters of *Jatropha curcas* – Biological activities and potential applications. In: G.M. Gubitz & M. Mittelbach (Eds.), Biofuel and industrial products from *Jatropha curcas* (pp.160-166).Graz: Dbv-Verlag University of Graz.

In: Jatropha Curcas
Editor: Gregory Medina

ISBN: 978-1-63483-089-8

Chapter 5

GROWTH AND SEED YIELD OF *JATROPHA CURCAS* L. CULTIVATED IN ARID REGION OF TUNISIA

Ezzeddine Saadaoui[1,*], Nizar Tlili[2,3], Naziha Ghazel[1], Chokri Ben Romdhane[1], Saad Abdelkebir[4], Mohamed Grira[4] and Mohamed L. Khouja[5]

[1]Regional Station of Gabes – LGVR. National Institute of Research in Rural Engineering, Waters and Forests (INRGREF), University of Carthage, Tunisia
[2]Faculty of Sciences of Gafsa. University of Gafsa, Tunisia
[3]Laboratoire de Biochimie, Département de Biologie, Faculté des Sciences de Tunis,Université Tunis El-Manar, Tunis, Tunisia
[4]Regional Commission for Agricultural Development, Gabes, Tunisia
[5]INRGREF, University of Carthage, Tunisia

ABSTRACT

In December 2007, seeds of *Jatropha curcas* from two African and six American provenances were introduced in Tunisia for the first time. A plantation was established in Southern Tunisia (region of Gabes). All plants were irrigated with treated wastewater. In 2013 and 2014, height, canopy circumference, seed yield and percentage of aborted seeds were studied for six and seven-year-old *J. curcas* plants. In 2013, four fructification periods were registered.

The highest seed yield was observed for a Brazilian provenance (A4) and its average was 398 g/plant. The lower was for A8 (Suriname provenance) with 134 g/plant. In 2014, six fructification periods were observed, the highest height and seed yield were registered for A3 (Brazilian provenance); their values were 297.5 cm and 804 g seeds/plant/year, respectively, and the lower seed production was reported for a Brazilian provenance (A2; 317 g/plant). In winter, seed yield is high, but it is characterized by a high percentage of aborted seeds. Oil content of seeds harvested in August 2013 varied from 23.46% (A7) to 32.15% (A5).

* Corresponding author: Ezzeddine Saadaoui. E-mail: saad_ezz@yahoo.fr.

Keywords: *Jatropha curacs* L., seed yield, provenance, treated wastewater, Tunisia

1. Introduction

Physic nut, *Jatropha curcas*, a member of the Euphorbiaceae family, is a drought tolerant shrub widely distributed in tropical and subtropical regions [1]. It has enjoyed increased popularity in recent years, due partly to its ability to grow in degraded zones and under arid and semi-arid conditions which does not require much water, fertilizer and management, and has high oil yield [2, 3]. *Jatropha curcas* has been found in altitudes from sea level up to 1800 m. The crop likes warm climate with mean annual temperatures between 19 °C and 28 °C and a 10 °C as a mean daily minimum temperature of the coldest month (T_{min})), whereas seed yield decreases and the plant dies after severe frost [4]. The best soil water environment for *J. curcas* cultivation is -3.4 kPa; where volumetric soil water content is 10% [5]. In addition, sandy soil reduced the growth of both shoot and root systems more than sandy-loam or clay-loam soils; there was particularly high plasticity in root and shoot thickness, as well as shoot length [6], and growth was better in sandy-loam soils than in clay-loam soils, and under recycled urban wastewater irrigation compared to desalinated brackish water [2, 7]. *Jatropha curcas* productivity was greater under recycled urban wastewater irrigation irrespective of the type of soil. The higher readily available nutrient content such as P and K provided by recycled urban wastewater appear to be the key factors behind the greater productivity [2]. Indeed, Jatropha plants showed best growth results with an N level of 30 kg/hectare and a P level of 10 kg P_2O_5/hectare in the first year and 45 kg N/hectare and 20 kg P_2O_5/hectare in the second year. The highest yields were obtained by applying 60 kg/hectare of N and 30 kg/hectare of P_2O_5 [4]. *Jatropha curcas* has shown to improve its water use efficiency along mild to moderate stress keeping a reasonable growth under soil water conditions up to 30% of Soil Water Availability (SWA) and resisting severe drought with a good recovery capacity after re-watering [8]. *Jatropha curcas* is sensitive to salt stress; stem height of young plants is reduced by increased NaCl concentration, this reduction is 46% at 0.2% NaCl (4dSm-1) compared to control [9]. In Asia and Africa, where jatropha is an introduced species, only leaf spots with low plant impact were recorded in fungal herbaria databases. However, serious pathogen problems have increased with mass planting, particularly stem cankers and root rot pathogens; *Phytophthora* sp., *Rhizoctonia* sp., and *Colletotrichum* sp. are all implicated in the disease complexes [10]. The most insects attack the plant and cause damage to fruits, inflorescences and leaves, belong to the genus *Pachycoris* (Heteroptera: Scutelleridae) [11].

Jatropha oil can be used as alternative fuel and for making biodiesel which overcome energy crisis problem [12]. In 2008, more than 900,000 ha of Jatropha plantations have been identified worldwide of which 85% is in Asia, 13% in Africa and the rest in Latin America, and by 2015 Jatropha is expected to be planted on 12.8 million ha worldwide [13, 14]. Experts from different countries agreed to ensure a seed yield of 4 – 5 ton/hectare/year that would be reasonable for its commercial viability.

It was estimated that with average seed yield of 3.75 ton/hectare, oil content 30 – 35% and oil yield 1.2 ton/hectare *J. curcas* would be superior than the other oil seed crops like soybeans (US) and rapeseed (Europe) which are producing 0.38 ton and 1.0 ton oil/hectare, respectively [13].

Jatropha curcas seeds are rich in oil that is easily convertible into biodiesel. 100-seed weight ranged from 40 to 74.6 g, the oil content ranged from 27 to 50%, free fatty acid ranged from 2.79 to 4.96%, and the biodiesel yield ranged from 82 to 90% [15, 16]. The major fatty acids found in the seed were oleic acid (41.5–48.8%), linoleic acid (34.6–44.4%), palmitic acid(10.5–13.0%), stearic acids (2.3–2.8%), cis-11-eicosenoic acid and cis-11,14-eicosadienoic acids [17]. Indeed, petroleum fuel produces carbon dioxide ($C0_2$) which is a harmful substance in highest quantity than the other transesterified *J. curcas* oil [18]. Also, the use of biodiesel and its blends has minimal effect on the environment and reduces the consumption of nonrenewable energy sources when compared to the diesel fuel [19]. *Jatropha curcas* is the source of many important bioactive compounds with high industrial and medicinal values [17]; it has been used as a multi-functional plant for traditional medicine, bio-pesticide, land erosion control, live hedge, oil for lighting and soap making [1]. In pharmacology, *J. curcas* has antimicrobial, antioxidant, antifungal, antidiabetic and anti-inflammatory activities [17, 20].

In addition, its press cake is a suitable biosorbent for removing Zn (II) in heavy metal polluted wastewater [21]. Some genotypes of *J. curcas* are toxic, due to the presence of alkaloids, known as phorbol esters, which cause a laxative effect, stomach pain and other discomfort on humans [22].

In order to evaluate the production of energy crops such as *Jatropha curcas* in severe conditions, eight provenances of *J. curcas* are introduced in the south of Tunisia, planted on marginal soil and irrigated with treated wastewater. The present study evaluates the potential for *J. curcas* production under irrigation with non-conventional water resources in marginal soil and arid climate. *Jatropha curcas* development and productivity are compared during two years, after six and seven years of plantation. In the first five years, all the eight provenances of this plantation showed a low seed production, the maximum is 256 g / plant / year for A4 (a Brazilian provenance) for five-year-old plants [23].

2. Experimental Site

The study was established at the regional station of National Institute of Research in Rural Engineering, Waters and Forests in Gabes (INRGREF) located at the South of Tunisia (33°54' N and 10°02' E), with an altitude of 44 m and the annual average precipitation, temperature and Potential Evapo-Transpiration (PET) are 130 mm, 24.5 °C and 1400 mm, respectively. This area belongs to the lower semi-arid bioclimate.

Eight provenances were studied, six from American continent: Brazil (A2 – A6) and Suriname (A8) and two from Africa: Mozambique (A1) and Tanzania (A7). Twenty five seedlings per accession were planted in December 2007. The field area is 1000 m^2, the spacing was 2 m x 3 m. (density: 1650 plants / hectare).

The soil is calcimagnesic with 15-35% of gypsum ($CaSO4\ 2H_2O$) at depths of 50-100 cm and 15% of calcium carbonate ($CaCO_3$).

The soil texture was drained and sandy; 85% sand, 10% clay and 5% silt. pH is 7.7 and electric conductivity EC = 2.2 mS/cm.

Jatropha curcas plants were irrigated with treated wastewater, characterized by pH = 6.9, EC = 4.670 mS/cm, COD (Chemical Oxygen Demand) = 168 mgO_2/l and BOD5

(Biochemical Oxygen Demand) = 40 mg O_2/l. The rate of Cl^-, Ca^{++}, Mg^{++} and SO_4^{2-} of treated wastewater is 39.4, 0.44, 0.32 and 450 mg × l^{-1} respectively. The average quantity of irrigation was 100 l/ week for a surface of 3 m^2.

3. Diversity of *J. Curcas*

Yue et al. [24] studied 276 accessions from five countries in South America; they did not detect any genetic diversity at 29 microsatellites loci. All accessions were homozygous at all loci and shared the same genotype at each locus, suggesting no microsatellite variation in the genome of *J. curcas*. But, DNA-marker polymorphisms and phenotypic traits of *J. curcas* indicated higher variability in Central American accessions compared to Asian, African and South American accessions [25].

For Montes et al. [26], genetic diversity in accessions from Central America and Mexico was higher than in accession from Africa, Asia, and South America. Other studies show limited genetic variation in *J. curcas* germplasm collections from South America, India, China, and Southeast Asia [27, 28]. In Mexico, the genetic study pointed out a high degree of similarity both within and among the non Mexican accessions. The Mexican accessions proved to be non toxic and genetically differentiated forming a well separated cluster from out of Mexico accessions.

Some polymorphic loci were closely correlated with the character toxicity and, once validated their association, useful in segregating populations for Marker Assisted Selection (MAS) [29]. *Jatropha curcas* can be improved through assessment of variation in wild sources and selection of superior/elite genotypes and application of mutation, alien gene transfer through inter-specific hybridization and biotechnological interventions to bring the change in the desired traits [30]. Our studies shows high phenotypic variability of *J. curcas*, essentially for leaf area, length, width and seed size and shape [23, 31, 32] and low molecular diversity [33]. Indeed, environment played significant roles in influencing phenotypic characteristics of *J. curcas* [34]. Molecular breeding should be primarily targeted to increase the seed yield, oil content, drought tolerance, abiotic stress tolerance, pest and disease resistance and also for developing optimum sized varieties for easy harvesting and management [35].

4. Growth of *J. Curcas* Plants

In 2013 and 2014, we studied height of plant (cm) and circumference of canopy (cm). In 2013, the lowest height was recorded in A4 (203.7 cm) and the highest was in A5 (241 cm).

In 2014, the lowest height was registered in A4 (252.9 cm) and the highest in A3 (297.5 cm). In 2013 and 2014, the smallest circumference was in A2 (625.7 cm) and A4 (885 cm), respectively. The highest was 775.6 and 1109.4 cm for A1 in 2013 and 2014, respectively (table 1).

5. Periods of Fructification and Seed Yield

Results presented in table 2 show four periods of fructification in 2013 and six periods in 2014. These results appeared in correlation with plant vigor. Important development in height and canopy circumference was observed in 2014. In addition, in previous years and in the same farm, we observed only three fruiting periods for all provenances [23]. In this plantation, *J. curcas* was reported to shed its leaves in the winter season, from January to April. In 2013, the highest seed yield was observed for A4, with an average of 398 g/plant (0.657 ton/hectare). In contrast, the lowest level was for A8 (134 g/plant/year; 0.221 ton/hectare). The maximum seed yield was registered in an individual from A5, with 1410 g/plant/year (table 2). In 2014, the lowest and highest seed yields were registered for two Brazilian provenances (A2 and A3), having 317 and 804g seeds/plant/year (0.523 and 1.326 ton/hectare) respectively. For example for A3, the maximum of seed production was observed in August (69%).

The most productive plant belongs to the provenance (A3) with 1860 g/plant/year (table 3). The last provenance is characterized by the highest height in 2014. In fact, the seed yield obtained in Tunisia was low. Jatropha seed yields reported for different countries and regions of Africa vary widely and range from 0.1 t/hectare to 15 t/hectare/year [36]. In Senegal, in pure culture, without size and inputs, the level of performance achieved by the fourth and fifth years does not exceed 200 kg / ha in the best cases [37]. In Mali, Abdoul Habou et al. [38] studied ten accessions; 211 kg/hectare was the best yield of seeds obtained for three-years-old plant of the best accession.

In Tunisia, Mediterranean conditions, seed yield of *J. curcas* is very low in the five years after production for eight accessions; the average of production varies from 200 to 400 g/plant [23].

Table 1. Average of height (cm) and circumference of canopy (cm) of each provenance in two years (2013 and 2014)

	2013		2014	
Provenances	Height	Circumference	Height	Circumference
A1	239.7	775.6	289.1	1109.4
A2	225.7	625.7	274.2	996.7
A3	225.9	722.1	297.5	990
A4	203.7	659.6	252.9	885
A5	241	701.7	283.5	1030.2
A6	231.2	721.1	288.9	982.85
A7	216.7	687.5	270.4	934.2
A8	233.8	710.5	265.7	987

Table 2. Average of seed yield (g/plant) of each provenance in 2013

Provenances	August 2013	October 2013	B-January 2014	End-January 2014	Maximum	Total/ individual
A1	63.6	29.3	102.85	27	368	175.4
A2	215.4	39.5	46	52	770	336.9
A3	199	26.7	61.6	58	1292	325.4
A4	256	69	53.4	36.6	1184	398.1
A5	144.4	34.7	133.4	39.2	1410	320.8
A6	211.3	43.6	30.4	46.3	570	322.9
A7	171.3	26.3	79.6	28	622	272.3
A8	9.3	16.7	74.9	33.5	546	134.5

In other conditions, seed yield was high; on two sites in Tamil Nadu, Southern India, the Jatropha variety (JO S2) produced up to 2.95 ton/hectare of dry seeds in the first year and up to 4.25 ton/hectare of dry seeds in the second year with 2 m × 2 m spacing between trees (a density of 2500 trees per hectare) [1]. The variation in seed yield is explained by endogenous factors: some genotypes are more productive than others; *J. curcas* possess some extent of diversity in days to flowering, fruit maturity, seed yield per plant and its component traits except number of seeds per fruit [39].

Also, main limiting factors of Jatropha production are soil conditions, altitude, climate (temperature, sunlight and rainfall extremities) and water logging conditions [7, 36, 40]. Saadaoui et al. [23] have studied the eight accessions from 2007 to 2012, and have registered high variability in production and large phenotypical [23, 31] and physiological variability [41].

6. Quality of Seeds

Figure 1 and 2 show important variation in percentage of aborted seeds. High percentage was observed in winter harvest for the two years (2013 and 2014). In the last harvest of 2013, the percentage of aborted seeds was between 61.5% for A8 to 81% for A5 (table 5). The lowest aborted seeds percentage was registered essentially in summer, the percentage of aborted seeds was less than 4.5% for A8 in 2013 and 9.8% for A2 in 2014. This result was observed in the field from 2007 to 2012 [31].

The low temperature of winter is the origin of the high percentage of aborted seeds, because, *J. curcas* is very sensitive to low temperature [42].

It does not tolerate frost, and flowers begin to appear only under specific temperature, radiation and phenological conditions [25]. In Tunisia, the region of Gabes which is characterized by a high winter temperature, because of its location in the Golf, is a favorable site for J. curcas production.

Table 3. Average of seed yield (g/plant) of each provenance in 2014

Provenances	August 2014	Beginning-September 2014	End-September 2014	End-October 2014	Mid-November 2014	December 2014	January 2015	Maximum	Total/ individual
A1	172	26.6	46.3	39.7	46.6	20.7	8	694.8	359,9
A2	174	37.5	36	18.3	22.8	13.9	14.5	730.6	317
A3	556	72.5	61.2	20.8	26.3	53.5	14.1	1850.6	804,4
A4	315	41	35.9	20.9	17.9	10.6	16	1236.6	457,3
A5	196	20	54.4	20.3	42.1	18.1	9.5	911.8	360,4
A6	391	23.7	28.4	12.5	13.8	16.9	8.3	1758.5	494,6
A7	307	45	49.7	66.2	60.2	23.5	21.3	794.4	572,9
A8	189	21.3	38.5	20.5	24.1	19.4	8.3	789	321,1

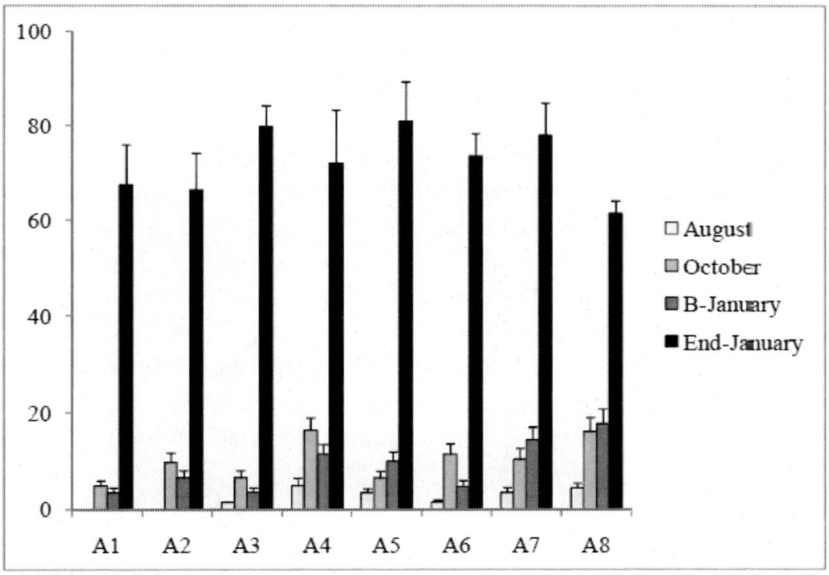

Figure 1. Percentage of aborted seeds in each period of fructification for eight studied provenances in 2013.

7. Yield of Seed Oil

The oil content of *J. curcas* seeds was determined according to ISO method 659:1998 [43]. Five grams of seeds were ground in a mortar and extracted with petroleum ether in a Soxhlet apparatus for 4 h. The solvent was concentrated on a rotary evaporator under reduced pressure at 60°C. The oil was dried by a stream of nitrogen and stored at -20°C until use.

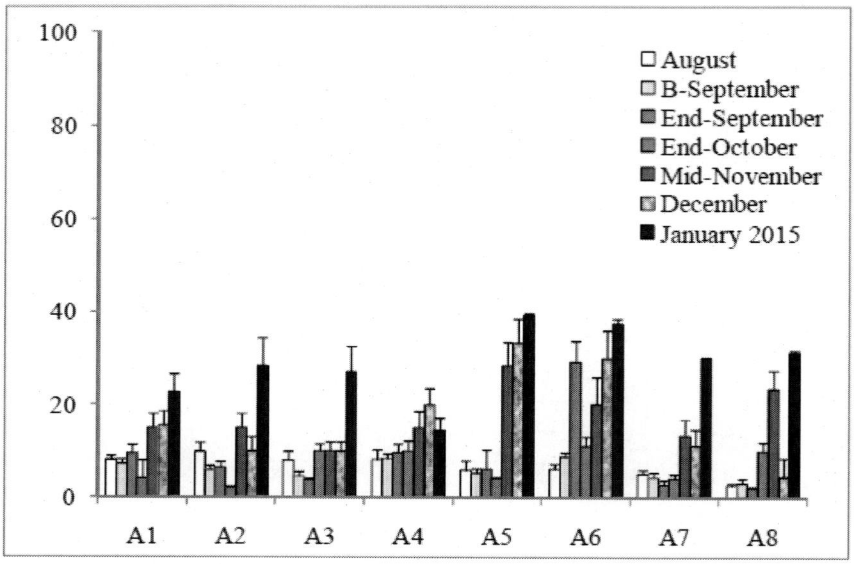

Figure 2. Percentage of aborted seeds in each period of fructification for eight studied provenances in 2014.

Oil content (table 5) varied between the eight studied provenances, the highest value was registered for A5 (32.15%) and the lowest for A7 (23.46%). All percentages were low in comparison with other results obtained in diverse countries. In Ghana, 44 to 49% oil content has been registered [15]. 46% was reported by Ugbogu et al. [3], in Nigeria. In Jharkhand, West Bengal and Bihar in Eastern India, Kumar and Singh [16] have studied 28 accessions and registered oil contents between 35 and 45%. Yi et al. [1] reported a crude fat percentage between 26.9 and 30.2%. Sushma [44] has reported a percentage between 30 to 35%. The oil content from "Congo-Brazzaville" variety ranged from 47 to 50% [45]. In India, Singh et al., 2014 have suggested a levels from 30 to 35%, Waghmare and Naik [46] registered seed oil content of 32.5%.

For these authors, the iodine value of the *J. curcas* seed oil was 348 mg/g which is higher than the other biofuel oil, the high iodine value due to its high content of unsaturated fatty acids. For Mohapatra and Panda [47], the seed oil content varied significantly under various regimes of N:P:K applications. Treatment with N_{50} P_{100} K_{60} and N_{60} resulted in consistent higher yield of seed oil.

Montes et al. [48] found strong effects of parental components and genetic pools on most seeds traits (oil content, seed mass, kernel mass, and the ratio of kernel to seed mass). But, the effect of self- and cross-fertilization on the oil content was not consistent as can be seen from the distribution of self- and cross-fertilized genotypes over the total range of oil content.

The genes related to polyunsaturated fatty acid synthesis were found to be down-regulated in leaves under drought stress, the proportion of polyunsaturated fatty acids (C16:3, C18:2, C18:3) in drought treated leaves was significantly lower than that in control leaves, while saturated fatty acids (C16:0, C18:0) showed the opposite pattern [49]. For Montes et al. [50], *J. curcas* oil is highly unsaturated, being linoleic acid is the most important component (42,6% to 53,3%).

Table 4. Oil content of seeds harvest in 2013 for each provenance

Provenances	Oil content (%)
A1	26.85 ± 1.2
A2	25.73 ± 0.9
A3	28.98 ± 0.6
A4	25.45 ± 0.7
A5	32.13 ± 0.2
A6	30.28 ± 0.5
A7	23.46 ± 1.3
A8	24.12 ± 0.2

In Morroco, Mokhtari et al. [51] obtained 13.68% of palmitic acid (13.68%), stearic acid (22,06%), oleic acid (20.46%) and linoleic acid (43,80%). The properties such as density, kinematic viscosity, flash point, pour point, acid number and moisture content of *J. curcas* biodiesel meet the international standard requirements of biodiesel [52].

CONCLUSION AND PERSPECTIVES

Jatropha curcas introduced to Tunisia shows high agronomic and phenotypical variability between plants and between provenances. The difference concerns morphologic traits, height, canopy circumference, seed yield, percentage of aborted seeds and oil content. Seed yield obtained in 2014 for the Brazilian provenance A3 (average seed yield 1.326 ton/hectare) suggest the possibility of *J. curcas* culture in arid region of Tunisia for the valorization of treated sewage effluent in energy crops. The selection of efficient genotypes, the use of selected genotypes, the introduction of other productive accessions from other countries characterized by heterogeneity and high productivity and the application of fertilization and pruning can significantly improve seed and oil yields of *J. curcas*. Also, some *J. curcas* genotypes are toxic; it is obligatory to find non-toxic plants before mass cultivation of *J. curcas*.

REFERENCES

[1] Yi, Ch., Reddy, Ch., Varghese, K., et al. A new *Jatropha curcas* variety (JO S2) with improved seed productivity. *Sustainability*, 2014, 6, 4355-4368.

[2] Dorta-Santos, M., Tejedor, M., Jiménez, C., et al. Recycled Urban Wastewater for Irrigation of *Jatropha curcas* L. in Abandoned Agricultural Arid Land. *Sustainability*, 2014, 6, 6902-6924. doi:10.3390/su 6106902.

[3] Ugbogu, A. E., Akubugwo, E. I., Uhegbu, F. O., et al. Quality assessment profile of *Jatropha curcas* (L) seed oil from Nigeria. *International Food Research Journal*, 2014, 21(2), 735-741.

[4] Sutterer, N. Jatropha cultivation using treated sewage effluent water requirements and environmental risks (a case of Southern Morocco). *Thesis of Master of Science degree*

in Agricultural Sciences. Faculty of Agricultural Sciences. University of Hohenheim, 2010, 60 p.

[5] Matsumoto, H., Yeasmin, R., Kalemelawa, F., et al. Effect of Soil Water Environment and Simulated Savanna Climate on Growth and Mineral Nutrition in *Jatropha curcas* L. *Agricultural Science*, 2014, 2(3), 13-22.

[6] Valdés-Rodríguez, O. A., Sánchez-Sánchez, O., Pérez-Vázquez, A., et al. *Jatropha curcas* L. Root Structure and Growth in Diverse Soils. *The Scientific World Journal Volume*, 2013, 9p. http://dx.doi.org/10. 1155/2013/827295.

[7] Rajaona, A. M., Sutterer, N., Asch, F. Potential of waste water use for Jatropha cultivation in arid environments. *Agriculture*, 2012, 2, 376-392. doi:10. 3390/agriculture2040376.

[8] Sapeta, H., Costa, M., Lourenc, T., et al. Drought stress response in *Jatropha curcas*: Growth and physiology. *Environmental and Experimental Botany*, 2013, 85, 76-84.

[9] Elhag, A. Z., Gafar, M. O. Effect of Sodium Chloride on Growth of Jatropha (*Jatropha curcas* L.) Young Transplants. *Universal Journal of Plant Science*, 2014, 2(1), 19-22.

[10] Ellison, C. A., Sawadogo, A., Braman, S., Nacro, S. First report of Colletotrichum truncatum causing stem cankers on *Jatropha curcas* in Burkina Faso. *Plant Dis*, 2015, 99, 14-20.

[11] Abdoul Habou, Z., Haubruge, E., Adam, T., Verheggen, F. J. Insectes ravageurs et ropriétés biocides De *Jatropha curcas* L. (Euphorbiaceae): synthèse bibliographique. *Biotechnol. Agron. Soc. Environ*, 2013, 17(4), 604-612.

[12] Sahay, S., Shyam, M. Storage conditions to improve the shelf life of *Jatropha curcas* seeds in terms of quality of oil. *J. Fundam. Renewable Energy Appl*, 2014, 4, 136. doi:10.4172/2090-4541.1000136.

[13] Singh, K., Patra, D. D., Singh, B., Verma, S. K. *Jatropha curcas*: A ten year story from hope to despair. *Renewable and Sustainable Energy Reviews*, 2014, 35, 356-360.

[14] Soliman, W. M., He, X. R. The Potentials of Jatropha Plantations in Egypt: A Review, *Modern Economy*, 2015, 6, 190-200. http://dx.doi.org/10. 4236/me.2015.62016.

[15] Derkyi, N. S. A., Sekyere, D., Oduro, K. A. Variations in oil content and biodiesel yield of *Jatropha curcas* from different agro-ecological zones of Ghana. *International Journal of Renewable and Sustainable Energy*, 2014, 3(4), 2014, 76-81. doi: 10.11648/j.ijrse.20140304.11.

[16] Kumar, S., Singh, S. Variability assessment of seed traits in *Jatropha curcas* L. for improvement of oil yield. *International Journal of Genetics and Molecular Biology*, 2014, 6(1), 8-15.

[17] Dubey, G., Bharati, K., Ahirwar, U., et al. Prospect of bioactive molecules from *Jatropha curcas* to improve soil and microbial quality for sustainable agricultura. *African Journal of Microbiology Research*, 2015, 9(3), 140-147.

[18] Lawrence, I., Jim, O. I., Uwaifo, A. J. Production of biodiesel from (RSO) rubber seed oil and *Jatropha curcas* seed oil (JCSO). *Standard Journal of Petroleum and Gas Exploration*, 2014, 1(1), 1-4. http://standres journals.org/journals/SJPGE/index.html.

[19] Kalaivani, K., Ravikumar, G., Balasubramanian, N. Environmental impact studies of biodiesel production from *Jatropha curcas* in India by life cycle assessment. *Environmental Progress and Sustainable Energy*, 2014, 33 (4), 1340-1349.

[20] Sundari, J., Selvaraj, R. *In vitro* evaluation of antibacterial and antifungal activity of *Jatropha curcas* L. *Rom. J. Biol. – Plant Biol*, 2012, 57 (1), 37-42.

[21] Abidin, Z. Z., Mohd Salleh, M. A., Harun, M. Y., Abu Bakar, N. Bioscrption of Zn (II) from aqueous solution by *Jatropha curcas* press cake. *Journal of Scientific and Industrial Research*, 2014, 73, 191-194.

[22] Rodríguez, O. A. V., Vázquez, A. P., Gamboa, C. M. Drivers and Consequences of the First *Jatropha curcas* Plantations in Mexico. *Sustainability*, 2014, 6, 3732-3746. doi:10.3390/su6063732.

[23] Saadaoui, E., Martín, J. J., Bouazizi, R., et al. Phenotypic variability and seed yield of *Jatropha curcas* L. introduced to Tunisia. *Acta Botanica Mexicana*, 2015, 110, 121-136.

[24] Yue, G. H., Lo, L. Ch., Sun, F., Cao, S. Y., Yi, Ch. X., Hong, X., Sun, W. B. No Variation at 29 Microsatellites in the Genome of *Jatropha curcas*. *Journal of Genomics*, 2014, 2, 59-63.

[25] Osorio, L.R.M., Salvador, A. F. T., Jongschaap, R. E. E, et al.High level of molecular and phenotypic biodiversity in *Jatropha curcas* from Central America compared to Africa, Asia and South America. *BMC Plant Biology*, 2014,14,77. doi: 10.1186/1471-2229-14-77.

[26] Montes, J. M., Technow, F., Martin, M., Becker, K. Genetic Diversity in *Jatropha curcas* L. Assessed with SSR and SNP Markers. *Diversity*, 2014, 6, 551-566. doi:10.3390/d6030551.

[27] Sun, Q., Li, L., Li, Y., et al. SSR and AFLP markers reveal low genetic diversity in the biofuel plant *Jatropha curcas* in China. *Crop Sci*, 2008, 48, 1865-1871.

[28] Rosado, T. B., Laviola, B. G., Faria, D. A., et al. Molecular markers reveal limited genetic diversity in a large germplasm collection of the biofuel crop *Jatropha curcas* L. in Brazil. *Crop Sci*, 2010, 50, 2372-2382.

[29] Vischi, M., Raranciuc, S., Baldini, M. Evaluation of genetic diversity between toxic and non toxic *Jatropha curcas* L. accessions using a set of simple sequence repeat (SSR) markers. *African Journal of Biotechnology*, 2013, 12(3), 265-274. doi: 10.5897/AJB12.2656.

[30] Divakara, B. N., Upadhyaya, H. D., Wani, S. P., Gowda, C. L. L. Biology and genetic improvement of *Jatropha curcas* L.: A review. *Applied Energy*, 2010.

[31] Bouaziz, R. Analyse de la variabilité phénotypique de huit provenances de *Jatropha curcas* introduites dans la region de Gabès. *Master in Biology and environment*, University of Gabes, 2012, 62 p.

[32] Bouazizi, R., Saadaoui, E., Ben Romdhane, Ch., et al. Variabilité inter-provenances de *J. curcas* introduite dans la région de Gabès. *Revue des Régions Arides*, 2014, 35(3), 255-260.

[33] Trebbi, D., Papazoglou, E. G., Saadaoui, E., et al. Assessment of genetic diversity in different accessions of *Jatropha curcas*. *Special Issue in the journal "Industrial Crops and Products" for the "Advances in Industrial Crops and Products Worldwide: AAIC 2014" international conference (in press)*, 2015.

[34] Wijaya, A., Surahman, M., Susantidiana, S., Lakitan, B. Genetic relationships among Indonesian *Jatropha curcas* L. accessions, selected crossings, and seed oil yield of their progenies. *Chang Mai J. Sci.*, 2014, 41(X), 1-12.

[35] Edrisi, S. A., Dubey, R. K., Tripathi, V., et al. *Jatropha curcas* L.: A crucified plant waiting for resurgence. *Renewable and Sustainable Energy Reviews*, 2015, 41, 855-862.

[36] Diaz-Chavez, R. Africa wide overview of jatropha production and bio energy. *Journal of Agricultural Science*, 2013, 1(4), 67-88.

[37] Terren, M., Cissé, C., Mergeai, G. Analyse des perspectives de rentabilité de la culture extensive de *Jatropha curcas* L. dans la zone agroécologique de transition orientale du Sénégal: cas de la communauté rurale de Dialacoto. *Cah Agric*, 2013, 22, 568-74. doi: 10.1684/agr.2013.0674.

[38] Abdoul Habou, Z., Katkore, B., Abasse, T., Verheggen, F. J. Évaluation du potentiel de dix accessions de *Jatropha curcas* L. (Euphorbiaceae) au Niger. *J. Appl. Biosci.*, 2014, 6456-6461.

[39] Aminul Islam, A. K. M., Anuar, N., Yaakob, Z., Osman, M. Heterosis for Seed Yield and its Components in Jatropha (*Jatropha curcas* L.). *International Journal of Plant Breeding* ,2011, 5(2),74-79.

[40] Taddese, H. Suitability analysis for *Jatropha curcas* production in Ethiopia - a spatial modeling approach. *Environmental Systems Research*, 2014, 3, 25. doi:10.1186/s40068-014-0025-7.

[41] Nasr, Z., Khouja, M. L., Aini, R., et al. Ecophysiological performances of eight *Jatropha curcas* provenances cultivated in Tunisia. *Journal of Applied Horticulture*, 2013, 15(1), 90-94.

[42] Najda, A., Almehemdi, A., Zabar, A. F. Chemical composition and nutritional value of jatropha *Jatropha curcas* L. leaves. *Journal of Genetic and Environmental Resources conservation*, 2013, 1(3), 221-226.

[43] ISO. International Standard ISO 659:1998. Oil seeds determination of hexane extract (or light petroleum extract), called “oil content”. ISO, Geneva, 1999.

[44] Sushma. Analysis of oil content of *Jatropha curcas* seeds under storage condition. *Journal of Environmental Biology*, 2014, 35, 571-575.

[45] Nzikou, J. M., Matos, L., Mbemba, F., et al. Characteristics and Composition of *Jatropha curcas* Oils, Variety Congo-Brazzaville. *Research Journal of Applied Sciences, Engineering and Technology*, 2009, 1(3), 154-159.

[46] Waghmare, T., Naik, G. R. Chemical analysis of seed and oil of Jatropha curcus biofuel plant cultivated in hyderabad Karnataka region, North Karnataka. *International Journal of Applied Biology and Pharmaceutical Technology*, 2015, 6(1), 186-192.

[47] Mohapatra, S., Panda, P. K. Effects of Fertilizer Application on Growth and Yield of *Jatropha curcas* L. in an Aeric Tropaquept of Eastern India. *Not. Sci. Biol.*, 2011, 3(1), 95-100.

[48] Montes, J. M., Bulach, A., Martin, M., Senger, E. Quantitative Trait Variation in Self- and Cross-Fertilized Seeds of *Jatropha curcas* L.: Parental Effects of Genotypes and Genetic Pools. *Bioenerg. Res.*, 2015. doi 10.1007/s12155-014-9576-8.

[49] Zhang, Ch., Zhang, L., Zhang, S., et al. Global analysis of gene expression profiles in physic nut (*Jatropha curcas* L.) seedlings exposed to drought stress. *BMC Plant Biology*, 2015, 15,17. doi: 10.1186/s12870-014-0397-x.

[50] Montes, J. M., Aliciardi, M. R., Chávez, J. V., et al. Characterization of *Jatropha curcas* L. seed and its oil, from Argentina and Paraguay. *The Journal of the Argentine Chemical Society*, 2011, 98, 1-9.

[51] Mokhtari, O., Hamdani, I., Chetouani, A., et al. Inhibition of steel corrosion in 1M HCl by *Jatropha curcas* oil. *J. Mater. Environ. Sci.*, 2014, 5(1), 310-319.

[52] Rahmath, A., Pogaku, R. Process optimization for biodiesel production from crude *Jatropha curcas* oil with immobilized lipase using response surface methodology. *J. Appl. Sci. and Agric.*, 2015, 10(3), 17-20.

In: Jatropha Curcas
Editor: Gregory Medina

ISBN: 978-1-63483-089-8

Chapter 6

PROPAGATION OF *JATROPHA CURCAS* THROUGH SEEDS, VEGETATIVE CUTTINGS AND TISSUE CULTURE

***A. K. M. Aminul Islam*[1,*], *A. K. M. Mominul Islam*[2], *Nor Anis Nadhirah*[3], *Nurina Anuar*[3] and *Zahira Yaakob*[3]**

[1]Department of Genetics and Plant Breeding, Faculty of Agriculture, Bangabandhu Sheikh Mujibur Rahman Agricultural University, Gazipur, Bangladesh

[2]Department of Agronomy, Faculty of Agriculture, Bangladesh Agricultural University, Mymensingh, Bangladesh

[3]Department of Chemical and Process Engineering, Faculty of Engineering and Built Environment, Universiti Kebangsaan Malaysia, Bangi, Selangor DE, Malaysia

ABSTRACT

Jatropha carcus is an important plant for the commencement of energy plantation. Jatropha can be grown in marginal, degraded, waste lands in tropical and subtropical regions of the world. It is a partially domesticated plant hence development of high yielding cultivars and package of management practices needs prime importance. The most important use propagation is to produce high quality planting materials. Jatropha can propagate through seeds, cuttings and tissue culture. Due to poor seed germination, short viability and high heterozygosity seed propagation may not produce high quality planting materials. On the other hand propagation through vegetative means or micropropagation can be the most effective alternatives to produce quality seedlings with desirable traits for higher seed and oil yield. Vegetative propagation has inimitable significance to maintain homogeneity among the progeny and provide early production.

* Corresponding author: A. K. M. Aminul Islam. Department of Genetics and Plant Breeding, Faculty of Agriculture, Bangabandhu Sheikh Mujibur Rahman Agricultural University, Gazipur 1706, Bangladesh. E-mail: aminuljkkp@yahoo.com.

The aspects of various propagation methods used in Jatropha cultivation such as seed propagation, vegetative propagation and micropropagation are highlighted in this chapter.

Keywords: Energy crop, micropropagation, cultivation, seed yield, biodisel

1.0. INTRODUCTION

Jatropha curcas L., a photo insensitive, drought tolerant, perennial plant belonging to the family Euphorbiaceae, emerges as a promising alternative for the future biodiesel (Openshaw, 2000; Pant et al., 2006).

Jatropha is an adaptable plant with a long history of cultivation in tropical Asia and Africa as a multipurpose plant (Catie, 2000; Henning, 2002, Yi et al. 2014). It is originated in Mexico where Jatropha oil has been utilized as raw materials for soap and fuel for lamps, different plant parts as medicinal purpose. It is universally accepted as an energy crop except few doubts. It also used as hedges around the home to avoid animals' and propagate through stem cuttings. Jatropha grows well in low fertile soil where little nutrient element is available. It can continue to exist in drought situation by dropping their leaves. Jatropha produce fruits within a year if cultivate properly. Depending on environmental conditions Jatropha plants produce seeds after approximately 2-3 years (Kobilke, 1989). It also produces fruits up to 50 years after plantation. If it planted in tropical climate, it is possible to harvest twice a year. Thus, Jatropha can be a suitable economic crop and offer income sources for the farmers. The number of seeds produces by Jatropha plant is very low but has profuse vegetative growth. Sexual propagation by seed directs genetic variability and makes the plants exposed to diseases. On the other hand, asexual propagation provides genetically homogeneous, disease-free planting material for commercial plantings (Nanda and Kochhar, 1987).

Detail studies on the propagation of Jatropha are inadequate and needs systematic study on different areas of its propagation.

1.1. Propagation

Propagation is the natural and artificial means of dispersal and transportation of seeds or plant parts to produce new plants. Propagation is defined as production of new individuals from a selected plant having all the characters of the original plant by means of seeds, cuttings, bulbs, corms, roots, leaves or tissue culture etc. (Brickell, 1992; Bhujbal, 2011). It is the method of increase of number of offspring's from mother plants.

Jatropha is usually propagated by seed. Healthy seeds are selected as a planting material for commercial cultivation. It is also propagated both by seed as well as by stem cuttings in case of large scale production. Availability of planting materials, propagation techniques, time of propagation and other required facilities should be considered for successful propagation. Direct seeding, transplanting, direct planting of stem cutting, or tissue culture techniques can be used for propagation (Freitas and Barjona, 1906).

There are two types of propagation methods (a) sexual propagation, and (b) asexual propagation or vegetative propagation (Aldriance and Brison, 2000). Plants often propagated by means of sexual reproduction through production of viable seeds.

When seeds do not germinate to grow into complete plant asexual propagation through cutting, grafting or tissue culture used to propagate plants.

Vegetative propagation permit plants to be propagated where plants fail to duplicate through sexual means (www.oliviassolutions.com).

1.2. Types of Propagation

1.2.1. Sexual Propagation

It is called sexual propagation because there is involvement of sexes and it occurs through fertilization of an egg by compatible pollen grains in the ovary of a flower. Sexual propagation produces seed which usually enclosed in a fruit. Mature seed germinates and begin active growth when it put into germination medium under favorable environment. Seeds are the principal way of sexual propagation (Bose et al., 1997). But it does not indicate every plant that produced seeds are sexually propagated.

Plant propagated from seed is comparatively simple in contrast to many techniques but it is sometimes difficult due to certain prerequisite of seeds germination (Sharma, 2002). Seed is an important plant parts which contain an embryo and have competent to grow in to whole plant. There are number of differences observed where plants produce apomictic seeds without sexual union of male and female gametes.

Seed propagation shows several advantages over other methods of propagations. Sexual propagation by seeds produces a large number of seedlings from a single plant (Aldriance and Brison, 2000). Seed can be used as a means of breeding new varieties.

It may store long time as a germplasm for future multiplication. Sexual propagation by means of seed needs a least quantity of materials and necessary understanding on seed biology. Species transmit their parental traits from parents to their progenies. Exploitation of heterosis is feasible only when the hybrids are propagated by sexual means in the first time and later fix heterosis through asexual propagation (Acquaah, 2007).

But the major disadvantage of seed propagation is the inconsistency of genetic material, not genetically true to type and limitation of saving seeds from hybrids for further propagation. Sexually propagated plant requires longer time to produce fruits. The plants can't be multiplied by this method which does not produce seed.

In self-pollinated plants each seed will bear the identical hereditary composition as the mother plant if there is absence of mutation. In cross-pollinated plants every seed contain a combination of genes of male and female parents.

On the other hand cross fertilization facilitates survival of a species in a varying environment by recombination of genes (Fryxell, 1957; Abrol, 2012). But seeds of cross pollinated plant like *Jatropha curcas* do not contain all desirable traits of the mother plant to breed true.

Jatropha seed comprises three most important parts: an embryo (miniature plant); the endosperm (stores food); and seed coat (protective cover) (Figure 1). High quality, healthy and viable seeds are essential for a successful propagation. Germination rate reflects the status of seed viability that means the higher the germination rate the higher the viability of seed.

Figure 1. *Jatropha curcas* (a, b) fruits in a cluster, (c) mature seed suitable for sexual propagation, and (d) cross section of seed showing embryo and endosperm.

1.2.2. Asexual Propagation

Asexual propagation is also known as vegetative propagation. In this method propagation is completed by cuttings, budding, grafting, or tissue culture. Asexual propagation produces genetically identical plants which retain the characteristics of parental plants.

It is also used to propagate plants when seed is tough to germinate or unavailable (Williams and Hanks, 1976). The plants propagated through asexual methods also propagated by seeds. The most important advantages of asexual propagation are the genetic uniformity, produce fruit within a short time, disease free plant. Once produce a desirable form of fruit, flower or plant, asexual propagation gives a means to reproduce that plant indefinitely (Duryea and Landis, 1984).

The major disadvantage of asexual propagation is the storage of genetic material. Storage is more complex to reproduce new plant from vegetative parts compared to the plant propagated by seed. The main difficulty of Jatropha cultivation is the reduced germination of seed (Islam et al., 2009).

On the other hand, vegetative propagation can save seeds and that seeds can be diverted for making biodiesel. Moreover, asexual propagation by cuttings or tissue culture produces true to type plants to their parents. Propagation through tissue culture can also be produced plants that can't be propagated by seed (Figure 2).

1.3. Significance of Propagation

Plant propagation is the procedure of proliferating plants of the same species to perpetuate their desirable traits (Sadhu, 1989).

Several other causes of practicing propagation are: (1) increases number of plants within a short time, (2) prevent variety or cultivar from extinction, (3) develop new varieties, and (4) produce disease free plantlets (5) seeds can be easily stored for long periods. Propagation material can improve by easy assessment from plants grown out of seeds.

The main purpose propagation is to produce high quality planting materials in case of Jatropha which maintain specific character for high seed and oil yield. High yielding plants usually have many branches and fruit bunches with more than 20 fruits per bunch (Islam et al., 2011). In case of Jatropha breeding, desirable stock can be selected from the variability present in seed propagated plants. Selected seeds can be used to develop new plants with desirable traits by removing poor performing plants from the seed stock.

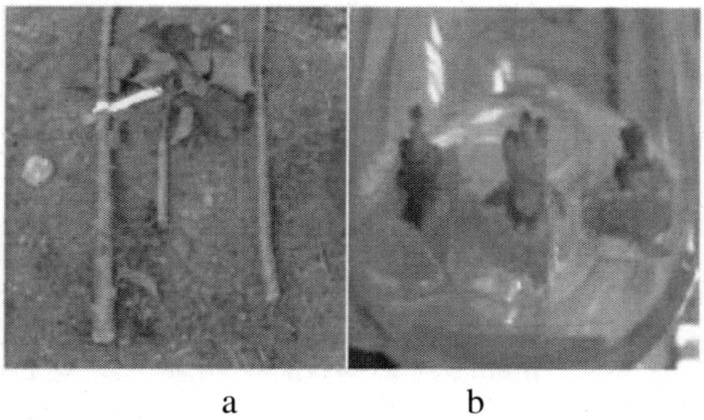

Figure 2. Asexual means of Jatropha propagation (a) stem cuttings (b) tissue culture.

1.3.1. Sexual Propagation

In sexual propagation seeds are produced through the process of pollination and fertilization of male and female gametes. In this method new plants are grown from seeds.

This is an essential method of genetic improvement of plants for qualitative and quantitative traits. Sexual propagation is important for the development of new varieties or hybrids. A large number of offspring's can be produced through sexual propagation within a very short time with minimum mortality (Poincelot, 1980). This method is quick, simple and economical in contrast to sexual propagation.

1.3.2. Asexual or Vegetative Propagation

In asexual propagation new plants are produced from vegetative parts of the mother plants such as stem cuttings, root cuttings, leaf cuttings, grafting or tissue culture. The new plants produced through vegetative propagation are the exact replicates of the mother plant since no fertilization are required and the progeny is the genetic copy or clone (Tchoundjeu et al., 1997; Waziri et al., 2015). The progeny contain all qualitative and quantitative traits of parental plant. Thus, selected plants can be propagated through asexual means without losing any traits of desire. Plant produce from this method can flower and fruit earlier than seed propagated plants. This method is used to propagate those plants which are difficult to propagate by seed or do not produce seed.

1.3.3. Micropropagation or Tissue Culture

Micropropagation is usually quicker compared to conventional methods of propagation and is independent of season. A large number of clonal plants can obtain from small explants within short time and space. For example, around 10,000 plants can be produced from single explants within 6-8 months (Soad and Mahdi, 2010). It has great importance in germplasm conservation where tissue or cells can be stored for longer time in small space. This method is useful where plants are difficult to propagate by seeds or by vegetative means and climates are not favorable to produce plant. Disease free plants can also be produced through tissue culture of meristematic explants. Tissue culture helps to develop novel genotypes for the improvement of new variety through induction of somaclonal variation. Other significant uses of plant tissue culture are haploid production, embryo rescue, somatic embryogenesis, in vitro mutagenesis, micrografting, genetic transformation, production of secondary metabolites (Beveridge et al., 1994; Croser et al., 2006; Mohamed et al., 2006; Patade and Suprasanna, 2008; Ribas et al., 2005).

1.4. Planting Techniques of Jatropha

Jatropha curcas can be cultivated by seeds, seedlings, vegetative cuttings or plant tissue culture (Figure 3). Seeds can be sown directly in the field or grown in polybags in the nursery to produce seedlings prior to transplanting in the field. Direct seeding is unfavorable due to low germination and high mortality of seedlings in establishing commercial plantation (Jabarzare et al., 2010). Jatropha cultivated from stem cuttings do not produce taproots, thus, plant grown from cuttings comprise short life span and susceptible to drought. The pseudo taproots produced from cuttings cannot enter into deep soil as taproots can.

A survey on 98 operational projects reported that 52% commercial project uses seedlings, 48% direct seeding and 29% vegetative cuttings for the establishment of Jatropha plantation (Wahl et al., 2012). Study also reported that vegetative cuttings are the excellent option for rapid production of flowers and fruits. On the other hand, plant establishment from seed is the best choice to produce high biomass.

1.5. Quality Planting Materials

Quality planting material is one of the important elements of successful agricultural production and development. Significant differences are found in yield and oil content in the plants grown from seed. Jatropha is largely propagated by seeds. Propagation through seed may not offer worthy planting materials for the establishment of sustainable plantation of Jatropha (Heller, 1996). Low viability and germination of seeds, high genetic variability makes the screening process laborious for quality seed. High genetic variability also negatively affects production of plantation. On the other hand the huge amounts of seeds are used as planting materials for raising plantation. Quality seeds for plantation should have 100% purity with no broken seed, more than 70% germination and less than 8% moisture.

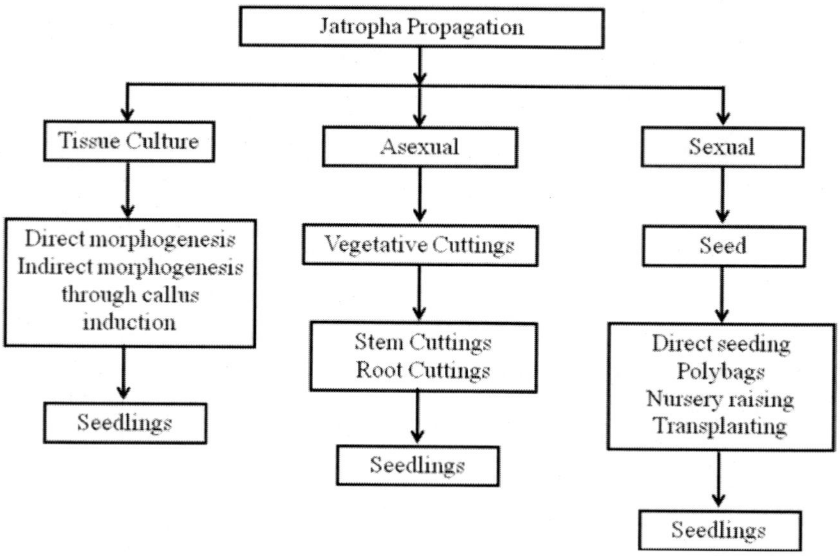

Figure 3. Propagation methods of *Jatropha curcas*.

Jatropha propagation can also be carried out without dropping the traits by vegetative cuttings (Islam et al., 2010; Resende et al. 2012) but the limitations of production of large scale planting materials are (a) availability of sufficient quantity of material, and (b) vegetative propagation is seasonal. The quality of planting material could be ensure by choice of good quality seed, polybags, propagation media, manure, humidity and pest control measure. Direct sunshine on the growing plants generates stiffness and resistance in the plant material.

2.0. Propagation by Seeds / Seed Propagation

Jatropha is generally propagated by seeds and incredibly unpredictable in terms of seed germination. Jatropha plantation can develop fast through seed propagation with no or very poor maintenance and attain at the height of three to eight meters. The fruits of Jatropha mature within 2-4 months after flowering. It is ready for harvest when it turns to yellow and the most suitable time when fruit color changed from yellow to black (Islam, 2011) (Figure 4). Seeds are split out from the fruit, cleaned and dried under shade to escape damaging effect on viability, germination if seeds planned to use for further plantation. Larger seeds should be selected for propagation and lighter seeds that do not sink into water should be discarded. Seeds should be dried to the moisture content of 5-7% before storing and stored in sealed jar. It is able to store at room temperature with maximum viability for one year. Seeds of *Jatropha curcas* retain better germination capacity in the storage up to six months after harvest. It generally shows low viability, delayed germination and root growth in the seedlings. Seeds are orthodox and lost their 50% viability within 15 months of storage (Kobilke, 1989; Islam, 2011). It is due to high oil content in the seeds. Warm and humid weather is favorable for better seed germination. It is essential to select best seeds with uniform size, color and discard damaged seeds from the high yielding plants for the production of quality seedlings. Seeds should not be dried under direct sunlight if want to use for propagation. Seeds should be dried properly at a temperature less than 25°C to maintain its germination quality. For getting good germination (80-95%) moisture content needs to be maintain 15-20% after drying. Seeds may show 70-75% germination six month after of storage when it dried up to moisture content of 7-8%. Seeds stored more than a year should not recommend for commercial plantation.

a b c d

Islam et al. 2009.

Figure 4. Showing Jatropha (a) fruit cluster in the plant (b) mature and green fruits, seeds, and (c, d) seedlings.

Viviparous germination is reported in Jatropha where seed germinate within the fruit before seed removed from fruit during rainy season (Deore and Johnson, 2008; Islam, 2011). There is evidence for seed dormancy just after harvest. Dry seeds are normally germinated without any pre-sowing treatments. Seed dormancy in many species occurs due to the presence of some chemical and physical inhibitors. The seed comprises a hard impermeable seed coat which stops imbibitions of water and germination. This impermeable testas exerts physical exogenous dormancy (Holmes et al., 1987). Hence, the seed necessitates pre-sowing treatments to achieve quick, regular and elevated germination (Demel and Mulualem, 1996). Seed dormancy can be broken by worm or cold scarification, mechanical or chemical scarification for softening seed coat to make it permeable to moisture. The common utilization of seed treatments before sowing is to hydrate seed which elevates rate of germination as contrast to directly sown seed without treatment.

Pre-sowing treatment improve seed germination by softening and breaking seed coat. It can be done by soaking in to water, scrapping to break hard seed coat, acid treatment to soften seed coat. Pre-soaking for 12 hours in cow dung slurry provided better seed germination (96%) compared to soaking in cold water or nicking the seed coat (72%) (Achten et al., 2008). Seeds that hydrated in to petridis with tap water and under sand stone for 72-hrs before sowing gave the maximum germination than directly sown seed, which is due to stimulus action of seed coat rupture (Feike et al., 2008).

Pre-sowing treatments can have various effects on different seed germination parameters of *Jatropha curcas*. Seeds placed on filter paper or kept under sand stone and moisten with water before sowing showed better performance for germination percentage, speed, index and seedling vigor index compared to unsoaked seeds placed for propagation (Islam et al., 2009). Seeds kept under sand stone and moisten was more effective than seeds placed on filter paper and moisten before sowing (Figure 5).

Untreated Jatropha seeds showed lower germination than those were treated before sowing (95.5%). Seed pre-treated in the stone sand gave the higher germination than untreated and other treatment (Table 1, Islam et al., 2009). Seeds kept under sand stone and moisten was better than seeds placed on filter paper and moisten for high percentage of seed germination, speed of emergence, energy of germination, germination index and seedling vigor index. Jatropha genotypes also acted differently to pre-sowing treatment (Islam et al., 2009). Selected seeds are soaked with water for eight hours before sowing. This is done to soften the seed coat for easy germination.

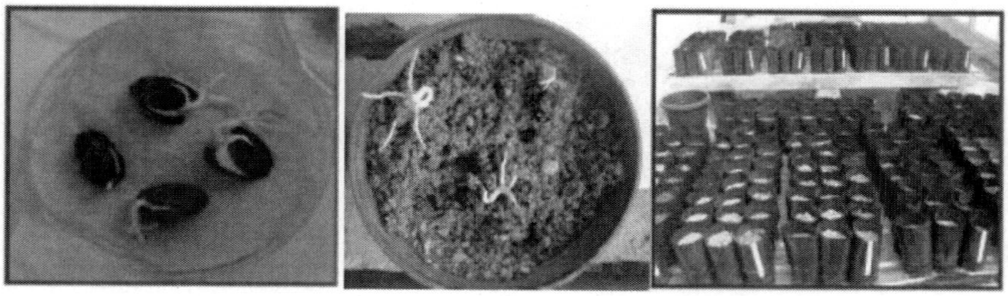

a b c

Islam et al. 2009.

Figure 5. Germination test of Jatropha seeds (a) moistened with water in petridish (b) seeds kept under stone sand and moistened with water, and (c) unsoaked seeds directly sown in the polybag.

Table 1. Effects of pre-sowing treatment on different germination parameters of *Jatropha curcas*

Pre-sowing treatment	GP	SE	EG	MGT	T_{50}	GI	SVI
Untreated or unsoaked seeds	77.10c	27.82c	44.12c	3.61c	7.14a	3.43c	14.08b
Seeds placed on filter paper and moisten with water	84.24b	36.59b	52.15b	3.72b	6.99b	3.50b	13.77c
Seeds kept under sand stone and moisten with water	95.85a	43.63a	59.55a	3.93a	6.98b	4.11a	15.59a

GP = Germination Percentage (%), SE = Speed of Emergence, EG = Energy of Germination, MGT = Mean Germination Time, T_{50} = Time of 50% germination, GI = Germination Index, SVI = Seedling Vigor Index, Means with the same letter are not significantly different.

Then sow seeds in the mixture of soil, sand and manure prepared in polybags or in the nursery bed or directly in the field with water. One seed per polybag should be sown at 2-4 cm depth with the upward orientation of seed chalazal end and downward orientation of seed caruncle (Srimathi and Paramathma, 2006). It is not suggested to take away seed coat prior to planting because it speedup germination but show possibility of receiving abnormal seedlings. The seed will germinate within 6-10 days after sowing (Islam et al., 2009). Water the polybags every day for 4-10 days and after germination every two days for the subsequent periods.

2.1. Direct Seeding

The easiest and cost effective methods for Jatropha plantation is the direct seeding in the field (Figure 6). Direct seeding method is popular to most of the farmers since they plant their crops in field in the same way. The rainy season is the best time for direct seeding in the field. Seed stored for at least one month to overcome seed dormancy is better to used as planting materials for direct seeding. It is observed that plantation established by direct seeding perform better in some areas and not good for other areas. The seeds should be soaked eight hours in water before planting at the depth of 4-6 cm. Since 1300 seeds weigh around one kilogram, one hectare of land at the spacing of 2 m x 2 m (2500 plants) requires around two kilogram seeds (Islam, 2011). Direct seeding requires more seeds than their normal seed rate. Rain is essential for proper seed germination and growth of the seedling. Generally, direct seeding requires no hole preparation, which results poor root development and affects the development of plant for higher yield.

2.2. Seeding in Polybags

Seeds planned for seeding in the polybags imperative to be soaked in water for eight hours prior to sowing. This is done to make softer the seed coat to assist quicker germination. Depending on local availability, polybags can be filled with soil, sand and manure or compost (1:1:1) or other mixtures. The mixture should be contained organic matter with free draining growing media (such as 1:1:1 sand-soil-manure or 1:1:2 sand-soil-compost).

The polybags needs to be well watered before sowing of seeds (Achten, 2008). Seeds are expected to germinate within 4-10 days after sowing (Islam, 2009). Sometimes germination of seeds in polybags is poor because of poor drainage and inadequate watering. The polybags needs very well aerated soil mixture with good drainage and judicious nutrient level. It must be adequately long to ensure proper taproot growth and development.

Polybags are watered daily before germination but after germination they should be watered as and when necessary. In case of delayed transplanting polybags should be transferred different place to avoid unwanted root growth anchoring in to the soil. Seedlings grown in the polybags will be ready for transplanting in the field after 45-60 days when they reach at the height of 30-40cm (Figure 7).

2.3. Seeding in Nursery Beds Prior to Transfer into Polybags

Raised beds of one meter width are prepared with the mixture of sand, soil and compost leaving 50 cm drainage in between beds (Figure 8). River sand also used to prepare nursery beds to get good results. Seeds should be sown at the depth of 2-3 cm in the beds.

It is better to soak seeds in water for eight hours prior to sowing in order to soften the seed coat to increase the rate of germination. Then seeds should be sown 2-3 cm deep in soil and moderate to frequent irrigation should be applied until germination complete. The healthy seeds will germinate within 4-10 days after sowing. The seedlings can be transferred in to polybags after 20-30 days and leave them under light shade. Depending on the climate seedlings should be hardened one month prior to transplant or transport to other place.

3.0. Vegetative Propagation by Cutting or Grafting

The achievement crop production and performance of its future generation greatly depends on the way of propagation. Vegetative propagation is highly preferred for all desirable traits (fruit yield, seed yield, oil content etc.).

a b

Figure 6. Showing (a) seedlings (b) plant established in the field from direct seeding.

a b c d

Figure 7. Showing (a) polybags prepared with mixture of sand-soil-compost (b) seeds sowing in the polybags (c) seed germinated in the polybags (d) seedlings transplanted in the field.

a b c

Figure 8. Showing (a) seedlings in the nursery bed (b) seedlings transferred in to polybags (c) seedlings ready for transplanting in the field.

The main advantage of cuttings is that they are genetically similar with mother plants. Propagation by cuttings is more reliable than propagation by seeds. Plants from cuttings have the same properties of high yielding mother plant (Nielsen, 2010; Singh and Shetty, 2012). On the other hand, the main disadvantage of the plant which developed from cuttings do not produces taproot, only produces lateral root and cannot access nutrients, water into deeper soil zone. Therefore plants from stem cuttings have limited capacity to tolerate drought. Jatropha propagation by cutting is only suitable when the land is fertile with good sources of permanent irrigation facilities and absence of prolonged dry period. Propagation by cuttings is the fastest and the cheapest means of *Jatropha curcas* cultivation. Generally farmer prefers propagation through cuttings because this type of propagation requires less time to establish productive plantation. Vegetative propagation can be done through stem or root cuttings or through grafting.

3.1. Stem Cutting

Vegetative propagation of *Jatropha curcas* by stem cutting is an important method after seed propagation. *Jatropha curcas* produces very low number of seeds but have profuse vegetative growth. On the other hand, seeds of Jatropha lost 50% of their viability within 15 months (Islam, 2011). Propagation by seeds directs genetic variability and exposed the plants to diseases while propagation through stem cuttings offers to develop genetically identical, disease free planting material for big plantation (Nanda and Kochhar, 1987). Jatropha is extremely cross-pollinated thus plants generated by seeds are highly heterozygous and genetically different from mother plant. Cuttings can be directly planted in the main field. However cuttings established in the polybags followed by transplanting in the field provide better results.

Propagation by stem cuttings without hormone application can be the advantageous and effective method for *Jatropha curcas*. In addition, it facilitates to maintain heterosis fixed in the seedlings for long time with no apprehension of segregation as there is no more recombination. Stem cuttings are best made from the mature and thickest branches by saw or sharp bolo about 30 cm long. It is better to collect stem cuttings from at least eight month old Jatropha plant if feasible from 45-100 cm from the bottom of the stem. Cuttings serve as a direct source of planting material and should have short internodes with many eyes.

It must be put (10-20 cm) directly into the soil leaving 15 cm or more of branch above the soil in a shaded nursery beds and watered to saturation. Soil should be kept wet throughout the period thus the best time for cuttings is the rainy season. It can also be planted in the polybags containing soil : sand : compost (1 : 1 : 1) in a nursery. Generally, the first shoots come out after 3-4 weeks. Stem cuttings can be separated into three basic types:

3.1.1. Hardwood Cuttings

Hardwood cuttings are collected from inactive branches of the plant from the growth of earlier season (hardwood) with no signs of active growth, no leaves and do not bend easily (Figure 9). Hardwood cuttings are planted into the polybag or directly to the field for rooting in the spring or rainy season. Cuttings are generally 30 cm long or above in case of field planting. The reduced rooting potentials in hardwood may be due to lesser phenolic levels as reported by Hartmann and Kester (1990) in some plant species. Hard wood cuttings produced higher number of shoots per cuttings, leaves per cuttings and lower shoot length (Islam, 2011). The hardwood cutting was found to be the best type of cuttings when treated with IBA because it gave the best performance compared to semi-hardwood and soft-wood cuttings of *Jatropha curcas* (Noor Camellia et al., 2009).

3.1.2. Semi-Hardwood Cuttings

Semi-hardwood cuttings are generally collected from partially mature wood of the existing season's growth, just after a flush of growth when it becomes woody (Figure 9). This type of cutting usually collects during July to September (Dumroese et al., 2012).

Figure 9. Showing (a) different types of cuttings (b) cuttings after 15 days of planting (c) cuttings after 30 days of planting (d) cuttings after 45 days of planting (e) shoot growth after 15 days (f) 30 days (g) 45 days of planting (h) cutting failed to continue growth (i) plant establish in polybag.

Semi hard wood cuttings took lesser days to new bud opening and shoot development as well as higher proportion of sprouted and rooted cuttings (Islam et al., 2010; Islam, 2011). Semi-hardwood cuttings took lower days to opening of new bud and shooting (4.8 and 11.7 d, respectively) as well as higher percentage of sprouted and rooted cuttings of 100% and 98.5%, respectively. The exogenous plant hormones considerably influenced the emergence of shoots, number of shoots per cuttings, length of shoot (cm) and number of leaves per cuttings in semi-hardwood cuttings. The efficiency of plant growth regulators on rooting and sprouting of semi-hardwood cuttings are in the order of IBA > NAA > IAA (Kumari et al., 2013).

3.1.3. Softwood Cuttings

Softwood cuttings are collected from soft, succulent, new growth of woody plants both in deciduous and evergreen species (Figure 9). Shoots are appropriate for making softwood cuttings when they can be bent easily and when the oldest leaves are mature and the newest leaves are still small. For most woody plants, this stage occurs from May to July.

The stem wood of each cutting should be partially matured, but not yet woody. Softwood cuttings are used to propagate lots of species, generally bushes or shrubs. This cutting comprises leaves and a number of nodes. Softwood cuttings are placed in the media and maintained underneath moist condition to prevent loss of water from leaves due to transpiration. It can be maintained in greenhouse, poly tunnels, plastic tent or plastic hoop houses with misting nozzles. Softwood cuttings are very much effective for several plant species and produce root within short time.

Root development in the cuttings has a lot of commercial benefits as there are many plant species that are difficult to root in cuttings. In some species, initiations of adventitious root occurs without any treatment of hormone whereas other requires different growth hormones generally auxin (Syros et al., 2004). Vegetative propagation is possible without using any growth hormones which can be a cost-effective procedure of multiplication of *Jatropha curcas* (Islam et al., 2010).

Significant variability was found in shoot and root development of three physiological ages of cuttings (hard, semi-hard and softwood cuttings) (Figure 10). Semi-hardwood cuttings took lower days to open new shoots (11.7) as well as high percentage of sprouted and rooted cuttings (100% and 98.5%, respectively).

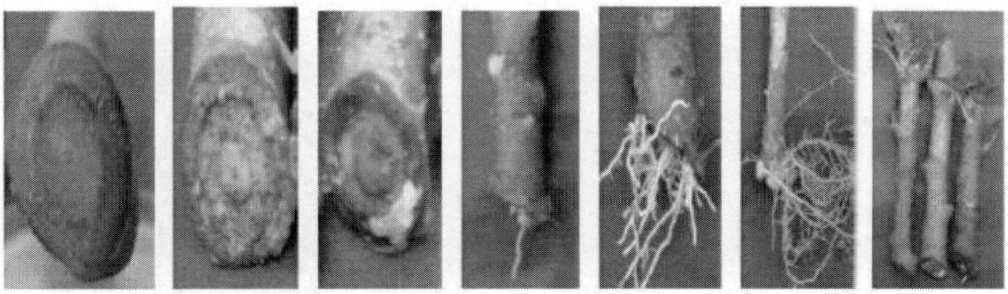

a b c d e f g

Figure 10. Development of adventitious root at the basal end of *Jatropha curcas* cutting (a) day 0 no changes occurred, (b) day 7 some callus developed at the basal end of cutting, (c) day 10 root primordial emerged from the callus (d) day 15 root emerged from the callus (e) root growth after 30 days (f) root growth after 45 days (g) shoot growth continue without root.

Soft wood cuttings took more days to open shoot and produced lower percentage of rooted cuttings. Semi hard wood cuttings proved to be more suitable for the vegetative propagation of *Jatropha curcas* through stem cutting, which gave more than 98% success. Genotypic differences were observed in shooting and rooting of Jatropha (Islam et al., 2010).

3.2. Root Cutting

Root cutting is an alternate process of vegetative propagation (Hartmann and Kester, 1990). A number of plants (rose, phlox, crabapple, fig, lilac, black berry, raspberry etc.) are propagated by root cuttings. Root cuttings of plants are generally collected during inactive stage (November to February) when levels of carbohydrate are high. Root cuttings of some species produce new shoots and afterward produce their own root system while root cuttings of other plants develop their root systems prior to generate new shoots (Dirr and Heuser, 1987). Roots are cut into pieces at least 5 cm in length and planted into polybags or pots containing soil : sand : compost in the ratio of 1 : 1: 1 or directly to the beds. During cutting a straight cut is made on the proximal end (nearest the crown of the parent plant) and a slanted cut on the distal end (furthest from the crown) of each root cutting. It is essential to maintain the accurate polarity of the cuttings. The distal ends should be inserted into the media leaving one third of the cuttings out of soil surface (Figure 11). Root cuttings is detrimental to the mother plant thus it is difficult to collect from the root system. Root cuttings do not necessitate high percentage of moisture condition during growth like stem cuttings.

3.3. Grafting

Jatropha curcas shows low seed viability, germination and delayed rooting of seedling as well as high degree of heterozygosity in the plantation (Kobilke, 1989; Islam, 2011). These constrains of seed propagated plants can be conquered by vegetative propagation through stem cuttings. Even though, multiplication by stem cutting is the fastest means of production of high yielding true to type Jatropha varieties, the stem cuttings are seasonal and shows lower longevity, posses poor tolerance to abiotic stresses compared to seed propagated plants (Heller, 1996).

a b c d

Figure 11. Showing (a) pieces of root cuttings (b) cuttings planted in the propagation media (c) shoot growth in the root cuttings.

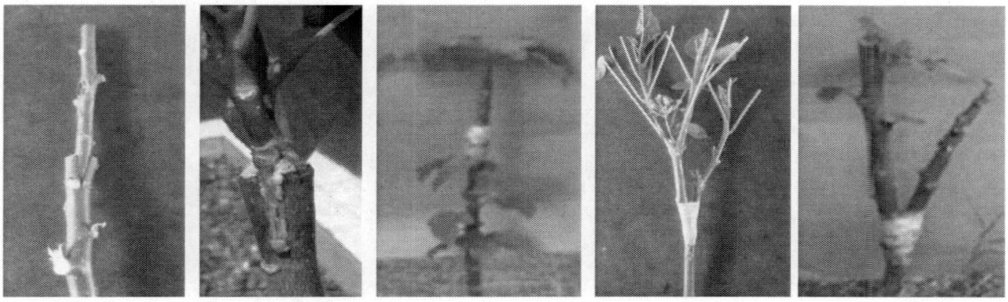

Figure 12. Grafting methods practiced for Jatropha propagation. (a, b, c) top cleft (d, e) side cleft grafting.

So propagation through grafting comes into sight to produce true to type, high yielding and disease free Jatropha plant. Grafting is the method of mounting a part of the plant (scion-part of the graft that produce the new system) on the root system (stock- lower part of the graft that posses a root system) of another plant (Figure 12).

Grafting is generally practiced as it is possess root system of plants tolerated unfavourable characteristics (growth, disease, etc.). Grafting method can be exploited in *Jatropha curcas* for the development of sustainable cultivation of at commercial scale.

A various types (top cleft, side cleft, veneer, whip, bark, contact and tongue) of grafting have been trying in *Jatropha curcas* during monsoon and spring to propagate high yielding genotypes (Panghal et al., 2013). Top cleft grafting method is found to be the most successful (> 90%), stable and strong amongst all grafting methods (Figure 12).

In Indonesia, *J. curcas* has been utilized as a root stock for *J. integerrima* for its gigantic growth. In India, grafting has been used to propagate hybrids of Jatropha (Panghal et al., 2013). The most perfect diameter of stock-scion was reported to be 3 to 4 cm and height of the grafting to be 40 to 60 cm. The success rate of grafting was higher in 3 to 4 cm scion-stock irrespective of grafting methods.

4.0. Propagation by Micro-Cutting or Mini-Cutting

Vegetative propagation through stem cutting, root cutting, grafting, budding, air layering has been widely used in nursery to maintain varietal purity. Rooting of cuttings is commonly complicated and there are many species which are recalcitrant for rooting in usual vegetative propagation (Klerk, 2002).

Utilization of micro-cutting has evolved as an accepted method for propagation of perennial plant species and has been established as an efficient and dependable method for propagation of desired plants (Titon et al., 2006). Micro-cutting permit to maintain genetic purity of mother plants (Wendling et al., 2010; Majada et al., 2011). The rooting ability of micro-cutting has been tested extensively in *Eucalyptus spp* (Almeida et al., 2007) and *Pinus pinaster* (Majada et al., 2011) with encouraging outcome.

Micro-cutting or mini-cutting is a technique where the juvenile part of the plants are cut into pieces and planted into the suitable media, so they can grow into a new plant (Awang et al., 2009).

In the recent times, micro-cutting method demonstrated a number of promises to propagate true to type plants in several horticultural crops. Microcuttings are generally done in a propagation chamber on a soilless medium containing of perlite 8 cm deep) underlying with light expanded clay aggregates (LECA).

4.1. Plant Materials or Shoot Cuttings

Juvenile top cuttings (4-6 cm in length having 2-3 nodes) from actively growing mother plants are used as plant materials in micro-cutting. Micro-cuttings can also be done from plants raised in micro-propagation under protected condition and grown in high planting density with regular pruning. Roots can be initiated from the cutting edge of stem and the number of adventitious roots varied significantly among genotypes or species (Awang et al., 2009). Day by day, micro-cutting technique is becoming popular and equally efficient for the propagation of different plant species. But this technique requires thorough investigation and perfection to increase rooting performance in *Jatropha curcas*.

4.2. Preparation of Mist-Chamber Propagation Box

Mist chamber is a useful means for the production of vigorous planting materials in agriculture. Awang et al. (2009) designed mist chamber propagation box for vegetative propagation by micro-cuttings in indigenous landscape plants in Malaysia. Mist chamber propagation box was built by using polystyrene box covered by polystyrene sheet (55 cm x 43 cm x 30 cm). This box is full of perlite (lava stone) at upper layer and light expanded clay aggregates (LECA) at the bottom level as propagation medium (Figure 13). The water is added and preserved before inserting micro-cuttings of stem at the base of the box. The water then flow through capillary pore to perlite. A small outlet or opening was set on the sidewall of the mist chamber at 5 cm above the base to maintain good air-water relationship in the root zone.

4.3. Insertion of Shoot Cuttings and Application of Treatments

The mist chamber propagation box is full with water up to definite level prior to placing of shoot cuttings into perlite. Afterwards, the box is covered with transparent polystyrene sheet which could allow light inside the box and prevent evaporation.

Distal end of micro-cuttings is inserted into the perlite. The optimum environment inside the mist chamber propagation box is the relative humidity: > 90%, temperature: 28.6-30.1C and light intensity: 673-2045lx. Appropriate temperature and humid environment (> 90% RH) can retain inside the mist chamber which helps for rooting of cuttings of different plants species and hardening of plantlets in tissue culture method. This technique is used to evaluate root formation during propagation of eucalyptus, bamboo, casuarine, stevia and many medicinal plants (Osman et al., 2013). Mist chamber propagation of *Jatropha curcas* can also be done by micro-cutting or mini-cuttings (Figure 14). More than 90% micro-cuttings produced root and continue growth to a whole plant (Islam, 2011).

Figure 13. Construction of mist chamber propagation box.

Formation of root is easier and faster in micro shoot cuttings or softwood cutting compared to semi-hard and hard wood cuttings. To accelerate uniform formation and increase the number of root, micro-cuttings can be treated with plant growth regulators.

Root formation in the cuttings depends on the size of shoot cuttings, plant growth regulators, percentage of light and shade (Osman et al., 2013). The treated micro-cuttings are inserted directly into perlite in the box in order to sprout roots. The micro-cuttings also inserted directly into perlite without being treated with plant hormones.

4.4. Hardening of Rooted Cuttings / Acclimatization

Rooted cuttings is transferred into the polybags containing mixtures of cocopeat, sand and soil (1:2:1) used as propagation media for hardening (Figure 15). Micro-cuttings then subjected to primary hardening inside poly tunnels with relative humidity of 90-100% for 30 days and partially opened poly tunnels with relative humidity of 65% for another 30 days before transplanting into the field. Secondary hardening done under shade of net with 45% relative humidity for another 30 days (Lavanya et al., 2009). Hardening of vegetatively propagated plants through micro-cuttings determine the superiority of the planting materials, the marketable production and the profitable feasibility of the nursery.

5.0. JATROPHA PROPAGATION BY TISSUE CULTURE

Plant tissue culture is a lab-based method which uses synthetic and sterilized media to propagate a large number of identical plants from small explants. Tissue culture is mainly utilized for multiplication of plant and is usually termed as micropropagation (Garg et al., 2011). It is also known as in vitro propagation, clonal propagation or express propagation (Assis et al., 2004). Plant tissue culture method necessitates sophisticated equipments and chemicals. That's why it is not suggested for small scale plantations. But it is important in nursery system to establish a tissue culture laboratory for large scale multiplication of planting materials. Different parts of the plant are used as explants in this method which permits propagation with desired traits of the mother stock (George et al., 2008). This method is important to tailor desirable traits (higher seed yield, oil content, early flowering and synchronous maturity) as true copy to a new variety. The advantage of plant tissue culture is that a large number of genetically identical planting materials can produce from a single plant with desirable traits within short time (Bhojwani and Dantu, 2013).

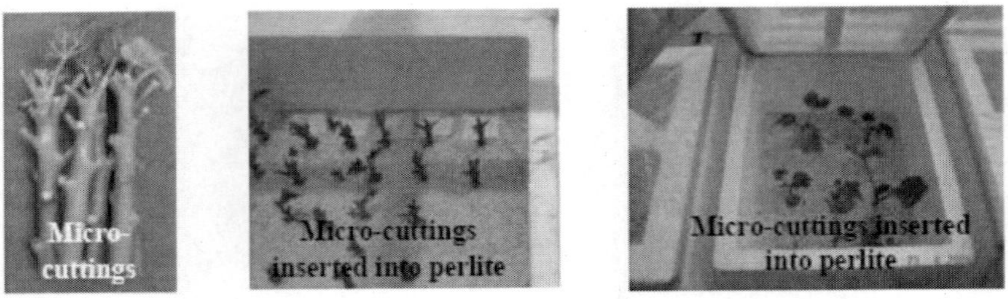

Figure 14. Preparation of microcuttings and insert them into mist chamber propagation box.

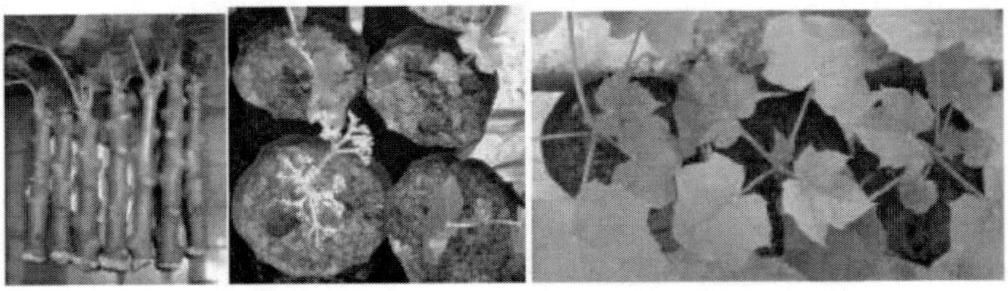

a b c

Figure 15. Transfer of (a) rooted cuttings from mist chamber propagation box to the (b) polybags (c) seedlings established in the polybags.

But the development of root system is not usual and grows vertically with the stimulation of hormones. Hundreds of plants can be produced from a small piece of tissue in a continuous process. Jatropha can be propagated by means of tissue culture and this method can use to develop high yielding new variety and conserved elite germplasm (Attaya et al., 2012). Tissue culture can also be promising methods of disease free Jatropha plant propagation because of its poor seed germination, scanty of rooting in cuttings (Purkayastha et al., 2010).

Micropropagation has immense prospects in large scale production of planting materials for commercial cultivation of Jatropha without conciliation of the quality (Heller, 1996; Yang et al. 2012). It is also important to produce secondary metabolites (alkaloids, terpenoids, propanoids etc.) from cultured cells. Genetic engineering along with tissue culture is very helpful for gene transfer.

5.1. Explants Sources for *J. Curcas* Tissue Culture

A number of efforts have been made to ascertain micropropagation protocols for Jatropha regeneration from different explants (Kumar and Reddy, 2010). The explants for Jatropha tissue culture are ususlly collected from three sources field, greenhouse or *in vitro* germinated seedlings. There are lots of variations in response of tissue culture from different explants (Yildiz, 2012). Growing conditions source plant and positions of organs of explats also affects tissue culture responses (Verma et al., 2013). Micropropagation of Jatropha has been achieved from different explants such as mature leaf, seeds, hypocotyles, juvenile cotyledon, petioles, axillary node through organogenesis and somatic embryogenesis (Sujatha et al., 2005; Jha et al., 2007; Shrivastava and Banerjee et al., 2008; Warakagoda and Subasinghe, 2009; Khemkladngoen et al., 2011; Panghal et al., 2012).

Nodal explants are the most excellent explants for micropropagation of Jatropha through tissue culture (Datta et al., 2007; Maharana et al., 2012) with minimum genetic variation (Pierik 1991). Explants from seeds are recognized more responsive for quick regeneration (Warakagoda and Subasinghe, 2009; Tiwari and Tuli, 2009).

5.2. Culture Media and Growth Ragulators

Murashige and Skoog's (MS) medium is reported to be the best culture media for all kinds of tissue culture in *Jatropha curcas* (Sujatha et al., 2005; Kalimuthu et al., 2007; Shrivastava and Banerjee, 2008; Kumar and Reddy, 2010; Kumar et al., 2010). B5 medium supplemented with BAP, Kn and NAA is also reported to use in Jatropha shoot proliferation (Warakagoda and Subasinghe, 2009). There are many types of growth regulators available with their own effect on growth and development of culture (Kumar et al., 2010).

Depending on the objectives of the culture, growth regulator and its concentration are used in tissue culture either separately or in a mixture. Plant growth regulators used in direct organogenesis includes indole-3-butyric acid (IBA), indole-3-acetic acid (IAA), 6-benzyl-amino-purine (BAP), kinetin (Kn), thidiazuron (TDZ) etc (Mukherjee et al., 2011).

5.3. Direct Organogenesis or Multiple Shoot Induction in Jatropha

The most reliable method of clonal propagation is the direct regeneration of shoot and root without intervening callus. The plants regenerated through direct organogenesis exhibits true genetic makeup of mother plant and show greater stability compared to callus induced plant (Mukherjee et al., 2011). In this method, explants were excised from shoot tip, node, petiole or hypocotyls of *in vitro* germinated seedlings and culture on MS media fortified with different combination of growth regulators (Sujatha et al., 2005; Shrivastava and Banerjee, 2008; Warakagoda and Subasinghe, 2009; Panghal et al., 2012).

Sujatha and Mukta (1996) successfully regenerated shoots from shoot tip culture using BAP and IBA in Jatropha for the first time. Kalimuthu et al. (2007) efficiently induced multiple shoot from shoot tip and nodal explants culture with combinations of BAP, kinetin and IAA. Benzyl-amino-purine (BAP) also found efficient growth regulator among other cytokinins in multiple shoot induction from shoot tip and nodal explants (Sujatha et al., 2005; Datta et al., 2007). Cotyledonary leaves and petioles produced higher number of shoot buds than the mature leaves and petioles (Kumar, 2009). It is also higher in horizontally placed petiole segments in culture media than vertically placed petioles (Kumar and Reddy, 2010).

The highest shoot regeneration observed from epicotyl explants of Jatropha when cultured in MS media supplemented with IBA (0.1 mgL^{-1}) and BA (0.5 mgL^{-1}) (Qin et al., 2004). Kalimuthu et al. (2007) achieved the highest number of shoots per explants (30 to 40 shoots per explants) within 30 to 40 days in Jatropha when cultured in MS media supplemented with BA (1.5 mgL^{-1}), Kin (0.5 mgL^{-1}) and IAA (0.1 mgL^{-1}).

In recent times, successful direct organogenesis (Figure 16) has been reported from various explants in Jatropha with or without auxins (IAA and IBA) and cytokinins (BAP and TDZ) where 100% regeneration was achieved (Kumar et al., 2011; Kumar and Reddy, 2010; Singh et al., 2010; Sharma et al., 2011; Kumar and Reddy, 2012).

Petiole a | Shoot tip b | Axillary bud c | Stem segment d | Nodal segment e

Source: Kumar and Reddy, 2013.

Figure 16. Direct multiple shoot regeneration from different explants.

Source: Kumar and Reddy, 2013.

Figure 17. Root initiation from *in vitro* regenerated shoot in half strength MS media supplemented with auxins.

Besides, regeneration efficiency mostly depends on genotype of the plant, source of explants, types of tissue or organ in Jatropha (Kumar, 2009; Mazumdar et al., 2011). Regeneration efficiency of *in vitro* explants is higher compared to *in vivo* explants (Sharma et al. 2011, Kumar and Reddy 2012). More shoot regeneration protocols for *Jatropha curcas* is available from Sujatha and Mukta (1996); Sardana et al. (2000); Lu et al. (2003); Wei et al. (2004); Sujatha et al. (2005); Rajore and Batra (2007); Jha et al. (2007), Mahalakshmi et al. (2014), Anjali et al. (2015).

5.4. Root Proliferation

Due to poor *in vitro* root proliferation it is difficult to propagate *Jatropha curcas* through micropropagation (Singh and Shetty, 2012). *Jatropha curcas* produces good number of multiple shoots during direct organogenesis but produces low percentage of roots in clonal propagation (Sahoo et al., 2011).

Another problem of micropropagation of plantlets is the failure or low success of establishment of root during acclimatization. Development of roots involved four different phases (i) elongation of shoots (ii) rooting of individual shoot (iii) satisfying dormancy needs of storage organ (iv) hardening of culture to enhance growth (Trigiano and Gray, 2005). The best rooting can achieve in half strength MS media supplemented with different concentrations of auxin (IAA, IBA, NAA) (Sujatha et al., 2005; Deore and Johnson, 2008; Singh et al., 2010; Anjali et al., 2015). Earlier investigation confirms that IBA is the most vital auxin for root development in *Jatropha curcas* and IAA and NAA are less effective (Kochhar et al., 2005; Attaya et al., 2012).

Kumar and Reddy (2010) found half strength MS medium is the best rooting medium fortified with 2% sucrose, 3.0 mg/l IBA along with 1.0 mg/l IAA, 1.0 mg/l NAA and 0.25 mg/l activated charcoal (Figure 17). Anjali et al. (2015) reported that 3.0 mg/l IBA along with 0.5 mg/l IAA in half strength MS media produces an average 6 to 8 roots per shoot. Rooting also achieved within 6 to 7 days in full strength MS media supplemented with 1.0 mg/l IAA (Kalimuthu et al., 2007; Attaya et al., 2012).

5.5. Embryo Culture of Jatropha

Embryo culture is use to test the vitality of seeds, production of rare species, production of haploids, rescue of embryos that abort on the mother plant after inter-specific or inter-generic hybridization (Randolph, 1945). Embryo rescue is ususally use to obtain viable plants especially where the production of embryo is often abnormal. The process begin when the unripe embryo is aborted in the F_1 generation to promote further development of embryo and then introduced in a suitable medium *in vitro*.

It is also used to shorten the breeding cycle by months or even years by breaking dormancy in seeds (Fathi and Jahani, 2012). Embryo rescue is important in order to interprete the associated factors affecting the dormancy and viability. The first reported embryo rescue in *J. curcas* was stated for further understanding of seeds dormancy and viability in *J.curcas* (Mohan et al., 2011).

5.6. Care and Handling of Tissue Culture Plantlets

Regenerated shoot with well developed roots are washed under tap water to remove remaining culture media before transfer in to sterile vermiculite pots. Plantlets transferred in to vermiculite pots are kept under poly tunnels shade with 75% RH, 25°C and 13 hrs daylight environments for two weeks for hardening. Shades of poly tunnel removed thereafter then transfer in to polybag containing soil, sand and compost in to the ratio of 1:1:1 (Figure 18). The well acclimatized seedlings are subsequently transferred to the field after two to three weeks. Augmentin could be used in the media during root proliferation to control bacterial contamination during hardening of *Jatropha curcas* plantlets (Toppo et al., 2012). The rooted plantlets then acclimatized in the soilrite under shade with 100% success.

6.0. Nursery Raised Plants

Nursery is the most essential components of any plantation program. It is a good alternative because the best plants can be selected where quality of seed and germination percentage of seed is poor. It is also possible to germinate larger amount of seeds directly in a seed bed then transfer to the polybags. The seedlings in the polybags can grow for 2 to 3 months before transplanting in the field. The seedlings grown directly in the seedbed are performing well if transfer without cutting their roots (Nielson, 2010; de Jongh and Nielsen, 2011).

Seedlings that show slower and abnormal growth can be removed in order to increase the quality and survival rate of planting materials in the nursery. The production also increase in nursery raised plants compared to the plants raised by direct seed sowing.

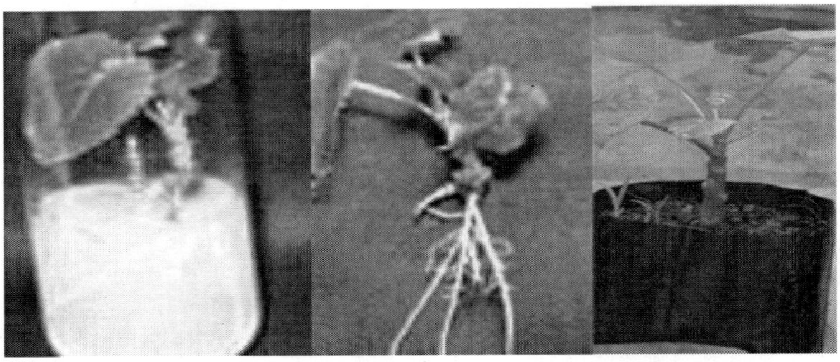

a b c

Source: Kumar and Reddy, 2013.

Figure 18. Seedling established from tissue culture (a) rooted cuttings (b) remove culture media before transfer to soil media (c) seedling established from *in vitro* culture.

There are four advantages of growing Jatropha seedlings in the nursery are (i) seedlings can be grown under control environment, (ii) abnormal seedlings can easily be discarded, (iii) nursery grown seedlings are stronger when planted in the field, and (iv) survival rates are high in the field after transplantation. The main weakness of nursery is hindered root growth of seedlings due to the size of polybags or container. This is normally happen when the seedlings cannot transplant timely in the field. There is also chance of infestation of insects and diseases to the seedlings and in the field at the time of planting. It may also involve extra labor and money in case of delayed planting in order to control diseases and pest. Usually big plantation requires nursery which must be equipped with (i) good source of water (ii) good source of propagation media (sand, soil, compost) (iii) sun or rain protection (iv) storage facilities (v) proper drainage (vi) proper transport facilities (vii).

7.0. CONSTRAINTS OF JATROPHA PROPAGATION

Traditionally, farmers' uses seed and stem cuttings for propagation during Jatropha cultivation. *Jatropha curcas* is self-compatible and produces seeds through cross pollination (Qing et al., 2007). Plants produce from cross pollinated seeds are heterozygous, not genetically true to type of the mother plant and requires longer time to produce fruits. Sexually produce Jatropha seed is overwhelmed with many troubles like poor germination, low viability, insufficient and slow rooting of seedlings.

Production begins in the seed propagated plant when it attains at 4-5 years old and production is still low. Propagation through vegetative cuttings is season dependent and produces plants with poor longevity. Vegetatively propagated plants are lack of tap root which make them susceptible to drought and disease (Sujatha et al., 2005).

On the other hand, plant tissue culture is expensive and requires skill manpower. The roots of tissue culture plantlets grow vertically with the stimulation of hormones but root growth is not usual.

CONCLUSION

Jatropha is a multipurpose plant produces seed containing 35-40% oil. It can grow under marginal soil with low rainfall condition. But the major problem of propagation is the poor and unreliable seed germination. Normal and plumpy seeds should be collected from fruits when the color turns to change from yellow to black to get maximum germination. Jatropha is propagated by seeds for large scale commercial plantation. Seed propagated plants are genetically heterozygous and cannot maintain its purity due to high cross pollination. Propagation through vegetative means can produce true to type offspring without losing any traits of the mother plant. It is also possible without using plant growth regulators in Jatropha. Large number of genetically identical planting materials can generate from a single plant within short time with desirable traits. This method can use to develop high yielding new variety, conserved elite germplasm and produce disease free seedlings. It has enormous potential in large scale production of planting materials. Development of high yielding variety and its propagation technology is the prerequisite for obtaining profitable yield of Jatropha in commercial cultivation.

REFERENCES

Abrol, D. B. 2012. Pollination for Hybrid Seed Production, In: *Pollination Biology-Biodiversity, conservation and Agricultural Production*. Springer Netherlands, pp. 397-411.

Acquaah, G. 2007. *Principles of Plant Genetics and Breeding*. Blackwell Publishing Ltd., John Wiley and Sons, Malden, US.

Aldriance, G. W., Brison, F. R. 2000. *Propagation of horticultural plants*. Mc Grow Hill Book Company. Inc., New York.

Almeida, F. D., Xavier, A., Dias, J. M. M., Paiva, H. N. 2007. Auxin (IBA and NAA) effects on mini-cuttings rooting of *Eucalyptus cloeziana* F. Muell. clones. *Revista Árvore*. 31(3): 455-463.

Anjali, K., Kumari, R., Shukla, L. N. 2015. Development of technique for efficient in vitro propagation of *Jatropha curcas* L. from nodal explants. *Botany*. 5(1): 40-43.

Assis, T. F., Fett-Neto de, A. G., Alfenas, A. C. 2004. Current techniques and prospects for the clonal propagation of hardwoods with emphasis on Eucalyptus. In: *Plantation Forest Biotechnology for the 21st Century*, C. Walter and M. Carson (eds.). Trivandrum-695 023, Kerala, India, pp. 303-333.

Attaya, A. S., Geelen, D., Belal, A. E. F. H. 2012. Progress in *Jatropha curcas* tissue culture, *American-Eurasian Journal of Sustainable Agriculture*. 6(1): 6-13.

Awang, Y., Rahman, N. A. and Sandrang, A. K. 2009. Adventitious rooting in microcuttings of selected indigenous landscape trees of Malaysia. *International Journal of Agriculture and Biology*. 11: 419-424.

Beveridge, C. A., Ross, J. J., Murfet, I. C. 1994. Branching mutant rms-2 in *Pisum sativum* (grafting studies and endogenous indole-3-acetic acid levels). *Plant Physiology*. 104: 953-959.

Bhojwani, S. S., Dantu, P. K. 2013. Micropropagation. In: *Plant Tissue Culture: An Introductory Text*, Bhojwani, S. S. and Dantu, P. K. (Eds.), Springer Netherlands, pp. 245-274.

Bhujbal, B. 2011. Unit 2: Plant Propagation Methods, *Resource Book on Horticulture Nursery Management.* Yashwantrao Chavan Maharashtra Open University, Nashik, India, 268 p.

Bose, T. K., Mitra, S. K., Sadhu, M. K., Das, P. 1997. *Propagation of Tropical and Sub tropical Horticultural Crops*. 2nd Edition, Naya Prakash, Calcutta.

Brickell, C. 1992. *Encyclopedia of Gardening*. Royal Horticulture society of London, Dorling Kindersley Publishers Limited, UK.

Croser, J., Lülsdorf, M., Davies, P., Clarke, H., Bayliss, K., Mallikarjuna, N., and Siddique, K., 2006. Toward Doubled Haploid Production in the Fabaceae: Progress, Constraints, and Opportunities. *Critical Reviews in Plant Sciences*, 25(2): 139-157.

Datta, M. M., Mukherjee, P., Ghosh, B., Jha, T. B. 2007. In vitro clonal propagation of biodiesel plant (*J. curcas* L.) *Current Science*. 93: 1438-1442.

de Jongh, J., Nielsen, F. 2011. *Lessons Learned: Jatropha for local development.* FACT Foundation, Wageningen, the Netherlands.

Demel, T., Mulualem, T. 1996. The effect of pre-sowing seed treatments, temperature and light on the germination of *Tamarindus indica* L., a multipurpose tree. *Journal of Tropical Forestry*. 12: 73-79.

Deore, A. C., Johnson, T. S. 2008. Occurrence of vivipary in *Jatropha curcas* L. *Current Science*. 95(3): 321-322.

Deore, C. A., Johnson, S. T. 2008 High frequency plant regeneration from leaf disc cultures of *Jatropha curcas* L.: an important biodiesel plant. *Plant Biotechnology Reports* 2, 7-11.

Dirr, M. A., Heuser, C. W. Jr. 1987. *The Reference Manual of Woody Plant Propagation: From Seed to Tissue Culture*. Varsity Press: Athens, Georgia.

Dumroese, R., Kasten, Landis, T. D., Luna, T. 2012. Raising native plants in nurseries: basic concepts. *Gen. Tech. Rep. RMRS-GTR-274*. Fort Collins, CO: US Department of Agriculture, Forest Service, Rocky Mountain Research Station. 84 p.

Duryea, M. L. and T. D. Landis (eds.). 1984. *Forest nursery manual: production of bare-root seedlings*. The Hague, M. Nijhoff/Dr. W. Junk Publishers, Oregon State University; Corvallis, OR, 385 p.

Fathi, H., Jahani, U. 2012. Review of embryo culture in fruit trees. *Annals of Biological Research.* 3(9): 4276-4281.

Feike, T., Mueller, J., Claupein, W. 2008. Examining germination rates of seeds of physic nut (*Jatropha curcas* L.) from Philippines and Viet Nam. Competition for Resources in a Changing World: New Drive for Rural Development. Tropentag, October 7-9, 2008, Hohenheim.

Fryxell, P. A. 1957. Mode of Reproduction of Higher Plants. *Botanical Review*. 23: 135-233.

Garg, P., Khatri, P., Gandhi, D. 2011. Plant tissue culture of *Jatropha curcas* L.: A review. *Imperial Journal of Pharmacognosy and Natural Products*. 1(1): 6-13.

George, E. F., Hall, M. A., Klerk, G. J., 2008. Plant Tissue Culture Procedure - Background. In: *Plant Propagation by Tissue Culture*, Volume 1. The Background, George, E. F., Hall, M. A., Klerk, G. J. (eds.), Springer Netherlands, pp. 1-28.

Hartmann, H. T., Kester, D. E., 1983. Plant Propagation. *Principles and Practices*, fourth ed. Prentice-Hall, Inc., Englewood Cliffs.

Hartmann, H. T., Kester, D. E., Davies, T. F. 1990. Principal of propagation by seed. In: *Plant propagation, principals and practices*. Chapter 6, pp. 104-136, Prentice-Hall International, Inc.

Holmes, R. J., McDonald, J. N. A. W., Juritz, J. 1987. Effects of clearing treatment on seed bank of the Alinene invasive shrubs *Acacia saligna* and *Acacia cyclops* in the southern and south western Cape, South Africa. *Journal of Applied Ecology*. 24: 1045-1051.

Islam, A. K. M. A. 2011. Improvement of biodiesel production through genetic studies of *Jatropha curcas*. *Ph.D. Thesis*, Department of Chemical and Process Engineering, Universiti Kebangsaan Malaysia, Bangi 43600, 254 p.

Islam, A. K. M. A., Anuar, N., Yaakob, Z. 2009. Effect of Genotypes and Pre-Sowing Treatments on Seed Germination Behavior of Jatropha. *Asian Journal of Plant Sciences*. 8: 433-439.

Islam, A. K. M. A., Anuar, N., Yaakob, Z., Osman, M. 2010. Propagation Potentials of Genotypes and Stem Cuttings of Different Physiological Ages of *Jatropha curcas* L. *Journal of Agricultural Science*, 2(4): 75-82.

Jabarzare, A., Bassiri, M., Yeganeh, H. 2010. Planting Methods and Seedling Establishment of Artemisia sieberi Besser: Seeds Collected from Isfahan Kolah Ghazi Sagebrush Vegetations. *Journal of Rangeland Science*. 1(1): 1-7.

Jha, T. B., Mukherjee, P., Datta, M. M. 2007. Somatic embryogenesis in *J. curcas* L., an important biofuel plant. *Plant Biotechnology Reporter*. 1: 135-140.

Jongh, J. 2010. *The Jatropha Handbook from Cultivation to Application*. FACT Foundation, Horsten 1, 5612 AX Eindhoven, The Netherlands, 15 p. (www.fact-foundation.com).

Kalimuthu, K., Paulsamy, S., Senthilkumar, R., Sathya, M. 2007. In vitro propagation of the biodiesel plant *Jatropha curcas* L. *Plant Tissue Culture and Biotechnology*. 17(2): 137-147.

Khemkladngoen, N., Cartagena, J., Shibagaki, N., Fukui, K. 2011. Adventitious shoot regeneration from juvenile cotyledons of a biodiesel producing Plant, *Jatropha curcas* L. *Journal of Bioscience and Bioengineering*. 111: 67-70.

Klerk, G. J. 2002. Rooting of microcuttings: theory and practice, *In Vitro Cellular and Developmental Biology*. 38(5): 415-422.

Kobilke, H. 1989. Studies on stand establishment of purging nut (*Jatropha curcas* L.). *Diploma thesis*. University Hohenheim, Stuttgart, Germany.

Kochhar, S., Singh, S. P., Kochhar, V. K. 2005. Effect of auxins and associated biochemical changes during clonal propagation of the biofuel plant *Jatropha curcas*. *Biomass and Bioenergy*. 32: 1136-1143.

Kumar, N. 2009. Studies on regeneration and genetic transformation in *J. curcas*. *Ph. D thesis*, Bhavnagar University, Bhavnagar, India.

Kumar, N., Reddy, M. P. 2010. Jatropha Tissue Culture: A Critical Review on Present Scenario and Future Prospects, In: *Jatropha, Challenges for a New Energy Crop: Volume*

2: Genetic Improvement and Biotechnology (eds. B. Bahadur et al. 2013), Springer Science, The Netherlands.

Kumar, N., Reddy, M. P. 2010. Plant regeneration through the direct induction of shoot buds from petiole explants of *Jatropha curcas*: a biofuel plant. *Annals of Applied Biology*. 156: 367-375.

Kumar, N., Vijay Anand, K. G., Reddy, M. P. 2010. In vitro Plant Regeneration of Non-toxic *Jatropha curcas* L.: Direct Shoot Organogenesis from Cotyledonary Petiole Explants. *Journal of Crop Science and Biotechnology*. 13(3): 189-194.

Kumari, A., Arya, M. C., Joshi, P. K., Ahmed, Z. 2013. Hormonal Effect on the Behaviour of Semi Hardwood Cuttings of *Jatropha curcas*. *The Indian Forester*. 139(2): 150-154.

Lavanya, M., Venkateshwarlu, B., Devi, B. P. 2009. Acclimatization of neem microshoots adaptable to semi-sterile conditions. *Indian Journal of Biotechnology*. 8: 218-222.

Mahalakshmi, R., Eganathan, P., Parida, A. 2014. In vitro regeneration from different ages of petioles of physic nut (*Jatropha curcas* L.). *African Journal of Biotechnology*. 13(2): 265-273.

Maharana, S. B., Mahato, V., Behera, M., Mishra, R. R., Panigrahi, J. 2012. In vitro regeneration from node and leaf explants of *Jatropha curcas* L. and evaluation of genetic fidelity through RAPD markers. *Indian Journal of Biotechnology*. 11: 280-287.

Majada, J., Martínez-Alonso, C., Feito, I., Kidelman, A., Aranda, I., Alía R. 2011. Mini-cuttings: an effective technique for the propagation of *Pinus pinaster* Ait. *New Forests* 41 (3): 399-412.

Mohamed, V. S., Sung, M. J., Jeng, L. T., Wang, S. C. 2006. Organogenesis of *Phaseolus angularis* L. high efficiency of adventitious shoot regeneration from etiolated seedlings in the presence of N6-benzylaminopurine and thidiazuron. *Plant Cell, Tissue and Organ Culture*. 86: 187-199.

Mohan, N., Nikdad, S. and Singh, G. (2011). Studies on seed germination and embryo culture of *Jatropha curcas* L. under *in vitro* condition. *Research Article, Biotechnology, Bioinformatics and Bioengineering*, 1,2,187-194.

Mukherjee, P., Varshney, A., Johnson, T. S., Jha, T. B. 2011. *Jatropha curcas*: a review on biotechnological status and challenges. *Plant Biotechnology Reports* 5, 197-215.

Nanda, K. K. and Kochhar, V. K. (1987). *Vegetative Propagation of Plants*. Principles and Practices, India Book Trust.

Nielsen, F. (2010) *Jatropha curcas* Oil Production for Local Development - Time to up scale?', *Technical report*, Banana hill, 23.

Osman, M., N. S. Samsudin, G. Faruq, and A. Nezhadahmadi. 2013. Factors Affecting Microcuttings of Stevia Using a Mist-Chamber Propagation Box. *The Scientific World Journal*, Volume 2013, 10 pages, Article ID 940201.

Panghal, S., Beniwal V. S. and Laura J. S. 2012. An efficient plant regeneration protocol from petiole explants of physic nut (*Jatropha curcas* L.). *African Journal of Biotechnology* Vol. 11(63), pp. 12652-12656.

Panghal, S., V. S. Beniwal and S. S. Soni. 2013. Generation of superior germplasm of *Jatropha curcas* through ex vitro grafting. *Agroforest System*, 87: 1023-1029.

Patade, Y. and P. Suprasanna, 2008. Radiation induced in vitro mutagenesis for sugarcane improvement. Nuclear Agriculture and Biotechnology Division, Bhabha Atomic Research Centre, Trombay, Mumbai, 400 085, Maharashtra, India. Published online: 1 June 2008.

Pierik, R. L. M., 1991. Commercial aspects of micropropagation, In: *Horticulture – New Technologies and Applications* (eds. J. K. Prakash and R. L. M. Pierik) Dordercht, The Netherlands.

Poincelot, R. P. 1980. *Horticulture: Principles and Practical Applications*. Englewood Cliffs, NJ: Prentice-Hall, Inc. US.

Purkayastha, J., T. Sugla, A. Paul, S. Solleti, P. Mazumdar, A. Basu, A. Mohammad, Z. Ahmed, L. Sahoo, 2010. Efficient in vitro plant regeneration from shoot apices and gene transfer by particle bombardment in *Jatropha curcas*. *Biol. Plant.*, 54: 13-20.

Qing, Y., P. D. Ping, D. Z. Biao, W. Z. Liang, S. Q. Xiang, 2007. Study on pollination biology of *jatropha curcas*, Euphorbiaceae. *South China Agric. Univ. J.*, 28: 62-66.

Resende, J. C. F., J. T. A. Silva, F. R. Simão, R. M. A. Pimentel, D. L. B. Morais. 2010. Phytotechnical Aspects of Jatropha Farming in Brazil, In: *Jatropha, Challenges for a New Energy Crop: Volume 1: Farming, Economics and Beyonds* (eds. N. Carles et al. 2012), Springer Science, The Netherlands.

Ribas, A. F., Kobayashi, A. K., Pereira, L. F. P., Vieira, L. G. E. 2005. Genetic transformation of *Coffea canephora* P. by particle bombardment. *Biol. Plant*. 49:493-497.

Sadhu, M. K. 1989. *Plant Propagation*, Wiley Eastern Limited, New Delhi, India.

Sahoo, N., Thirunavoukkarasu, M., Behera, P. R., Satpathy G. B., Panda, P. K. 2012. Direct shoot organogenesis from hypocotyl explants of *Jatropha curcas* L. an important bioenergy feedstock. *GCB Bioenergy*. doi: 10.1111/j.1757-1707.2011.01120.x.

Sharma, R. R. 2002. *Propagation of Horticultural crops: Principles and Practices.* Kalyani Publishers, New Delhi.

Shrivastava, S., Banerjee, M. 2008. *In vitro* clonal propagation of physic nut (*J. curcas* L.): Influence of additives. *Inter. J. of Integr. Biol.* 3: 73-79.

Singh, A., Reddy, M. P., Chikara, J., Singh, S. 2010. A simple regeneration protocol from stem explants of *Jatropha curcas* - A biodiesel plant. *Ind. Crop Prod.* 31 209-213.

Singh, G., Shetty, S. 2012. Evaluation in vitro rooting efficiency in the biodiesel plant, *Jatropha curcas*. Biotechnology. *Bioinformatics and Bioengineering*. 2: 591-596.

Soad, A. A. A. and S. M. Mahdi. 2010. Commercial production of tissue Culture date palm (*Phoenix dactylifera* L.) by inflorescence technique. *Journal of Genetic Engineering and Biotechnology*. 8(2): 39-44.

Sujatha, M., Makkar, H. P. S., Becker, K. 2005. Shoot bud proliferation from axillary nodes and leaf sections of non-toxic *J. curcas* L. *Plant Grow. Regu.* 47: 83-90.

Syros, T., Yupsanis, T., Zafiriadis, H., and Economou, A. (2004). Activity and isoforms of peroxidases, lignin and anatomy, during adventitious rooting in cuttings of *Ebenus cretica* L. *Journal of Plant Physiology*, 161: 69-77.

Tchoundjeu, Z., De Wolf, J. and Jaenicke, H. 1997. Vegetative propagation for domestication of agroforestry trees. *Agroforestry Today*. 9(2): 10-12.

Titon, M., Xavier, A., Otoni, W. C. (2006). Clonal propagation of *Eucalyptus grandis* using the mini-cutting and microcutting techniques. *Scientia Forestalis* 71: 109-117.

Tiwari, S., R. Tuli, 2009. Multiple shoot regeneration in seed-derived immature leaflet explants of peanut (*Arachis hypogaea* L.), *Scientia Horticulturae*, 121, 223-227.

Toppo, D. D., Singh, G., Purshottam, D. K., Misra, P. 2012. Improved in vitro rooting and acclimatization of *Jatropha curcas* plantlets. *Biomass and Bioenergy*. 44: 42-46.

Wahl, N., T. Hildebrandt, C. Moser, F. L. Freund, K. Averdunk, R. Bailis, K. Barua, R. Burritt, J. Groeneveld, A. M. Klein, M. Kügemann, D. Walmsley, S. Schaltegger, and T.

Zelt. 2012. Insights into Jatropha Projects Worldwide. Centre for Sustainability Management, Leuphana University of Lüneburg, Scharnhorststr.1, D-21335 Lüneburg, Denmark.

Warakagoda, P. S. and S. Subasinghe. 2009. In Vitro Culture Establishment and Shoot Proliferation of *Jatropha curcas* L. *Tropical Agricultural Research and Extension*. 12(2): 77-80.

Waziri, M. S., B. A. Kyari, M. Ibrahim, B. Apagu, B. Yunana, M. N. Askira, and A. B. Benisheikh. 2015. Effect of different soil media on the rooting and growth of *Delonix regia* stem cuttings in Maiduguri. *AshEse Journal of Agricultural Science*. 1(2): 014-019.

Wendling, I., Brondani, G. E., Dutra, L. F., Hansel, F. A. 2010. Mini-cuttings technique: a new *ex vitro* method for clonal propagation of sweetgum. *New Forests*. 39(3): 343-353.

Williams, R. D. and S. H. Hanks. 1976 (slightly revised 1994). Hardwood nursery guide. USDA Forest Service, *Agriculture Handbook* 473, 78 p.

Yang, C. Y., Z. Fang, B. Li, Y. F. Long. 2012. Review and prospects of Jatropha biodiesel industry in China. *Renewable and Sustainable Energy Reviews*, 16: 2178-2190.

Yi, C., C. Reddy, K. Varghese, T. N. H. Bui, S. Zhang, M. Kallath, B. Kunjachen, S. Ramachandran, and Y. Hong. 2014. A New *Jatropha curcas* Variety (JO S2) with Improved Seed Productivity. *Sustainability*. 6: 4355-4368.

In: Jatropha Curcas
Editor: Gregory Medina

ISBN: 978-1-63483-089-8

Chapter 7

ALLELOPATHIC EFFECTS OF AQUEOUS EXTRACTS OF *JATROPHA CURCAS* L. ON THE GERMINATION OF SIX CULTIVATED SPECIES

***Ezzeddine Saadaoui*[1,*], *José J. Martín*[2], *Naziha Ghazel*[1], *Chokri Ben Romdhane*[1], *Mohamed L. Khouja*[3] *and Emilio Cervantes*[2]**

[1]Regional Station of Gabes, Laboratory of Management and Valorization of Forest Resources, National Institute of Research in Rural Engineering, Waters and Forests (INRGREF), University of Carthage, Tunisia
[2]INRGREF, University of Carthage, Tunisia
[3]IRNASA-CSIC, Salamanca, Spain

ABSTRACT

The study concerns the effect of the aqueous extract of physic nut (*Jatropha curcas* L.) on the germination of six cultivated crops: *Hordeum vulgare* L., *Medicago sativa* L., *Corchorus olitorius* L., *Trigonella foenum-graecum* L., *Lens culinaris* L. and *Cicer arietinum* L. The extracts were obtained after an incubation of the dry plant material in distilled water (9%) during 48 hours at 60°C. The extraction was made from five organs of *J. curcas* (root, twig, leaf, seed and pericarp). The results showed a variable behavior between studied organs and species. *Corchorus olitorius* is the most tolerant species, showing elevated germination rate (> 90%) for all treatments. *L. culinaris* is the most sensitive species; its rate of germination is lower than 23% for all treatments with *J. curcas* aqueous extracts. The pericarp aqueous extract is the most inhibitive of the germination; we obtained rates of 0% for *T. foenum-graecum, L. culinaris* and *M. sativa*. The aqueous extracts of *J. curcas* affect also root length; a reduction of the parameter was observed for all studied species, essentially with the aqueous extract of pericarp. Allelochemicals may be present in any parts of *J. curcas* but fruit pericarps are the major source.

---

* Corresponding author: Ezzeddine Saadaoui. E-mail: saad_ezz@yahoo.fr.

Keywords: Allelopathy, *Jatropha curcas,* aqueous extract, germination, root length, cultivated species

INTRODUCTION

Allelopathy is defined as the capacity that a specific plant has of directly or indirectly interfering in another one's metabolism, by liberating chemical compounds elaborated by itself in the environment, influencing in plant domination, plant community formation and vegetation climax, as well as in the productivity and handling of cultures [1]. The chemical substances, collectively known as allelochemicals, are usually secondary plant products or waste products of main metabolic pathways of plants [2]. Commonly plants produce a large variety of secondary metabolites like phenol, tannins, terpenoids, alkaloids, polyactylene, fatty acids, steroids, which have allelopathic effects on the growth and development of the neighboring plants [3].

Allelochemical substances can be present in the leaves, bark, roots, root exudates, flowers and fruits. The delivery of allelochemicals into the rhizosphere is often thought to occur through leaching from leaves and other aerial plant parts, through volatile emissions by roots exudation and by the breakdown of bark and leaf litter [4]. The analysis of allelopathic effect of the aqueous extract of a species on the germination of other species is a method to estimate the allelopathy between species. *Jatropha curcas* (Euphorbiaceae), or physic nut, has straight trunk and grey or reddish bark, masked by large white patches. It has green leaves with a length and width of 6 to 15 cm, with 5 to 7 shallow lobes. The leaves are arranged alternately. Dormancy is induced by fluctuations in rainfall and temperature/light [5].

Yue et al. studied 276 accessions from five countries in South America; they did not detect any genetic diversity at all loci in 29 microsatellite. All accessions were homozygous at all loci and shared the same genotype at each locus, suggesting no microsatellite variation in the genome of *J. curcas* [6]. Jatropha, a multipurpose, drought resistant, has gained lot of importance for the production of biodiesel [7]. The *J. curcas* industry is in its very early stages, covering a global area estimated at some 900,000 ha in 2008, 760,000 in Asia, 120,000 in Africa and perhaps 20,000 in Latin America [8, 9]. *J. curcas* has been utilized historically for hedging and in green manure and fertilizers, food, soaps, pesticides, charcoal, and traditional medicine [10]. In pharmacology; it has antimicrobial, antioxidant, antifungal, antidiabetic, anti-inflammatory, antiulcer, hepatoprotective, antimetastatic, antiproliferative, wound healing, pregnancy terminating and anthelmintic activities [6, 11]. The fruit of *Jatropha curcas* can be a very good source for herbal drugs especially in cold aqueous, ethyl acetate and methanol as solvents [12]. Latex, twigs, roots and leaves are used to treat skin disease such as ring worm, scabies, eczema, and sexually transmitted disease [13].

Most varieties of *J. curcas* cultivated in the world are toxic; the phorbol esters found in the seed and oil were identified as the main toxic agents of *J. curcas*, animal consumption of these phorbol esters can cause diarrhea and inflammation of the gastrointestinal tract and death [14]. *J. curcas* has allelopathic effect inhibiting growth of different crops [15], its aqueous extracts of leaves and roots had high effect and inhibited the germination and growth of other species [1, 16, 17, 18, 19]. The presence of phenolic compounds could be the cause of reduction in germination [20].

Indeed, total phenolics in methanol extract of *J. curcas* leaf amounted to 35 $mg.g^{-1}$; kaempferol, coumarin, catechin and quercetin acids were detected in this extract [21].

The main allelopathic substance is azelaic acid. This compound inhibited the germination of seeds and the shoot and root growth of corn (*Zea mays* L.) and tobacco (*Nicotiana tabacum*). The percentage of azelaic acid in distilled water extracts of leaves was at least 0.94%, which indicated that azelaic acid may provide a competitive advantage to *J. curcas* in the defense mechanism by inhibiting the growth of neighboring crops [16].

This study aims to determine the effect of aqueous extract of *J. curcas* (leaves, twigs, roots, seeds and pericarps) on the germination of six cultivated species. Three of these species (*H. vulgare, M. sativa* and C. *olitorius*) showed a decrease of germination percentage when treated with aqueous extract (9%) of *Eucalyptus occidentalis, Acacia ampliceps* and *Prosopis juliflora* [22].

MATERIAL AND METHODS

1. Sample Collection and Extraction

Leaves, twigs, roots, seeds and pericarps of physic nut (*Jatropha curcas*) were collected from cultivated plants in the region of Gabès (South of Tunisia; 33°54' N and 10°02' E).

The aqueous extracts were prepared from dried plant material ground in powder form. Nine grams of powered *J. curcas* material were added to 100 ml distilled water and left for 48 h at room temperature (25-30°C). The extracts were filtered with filter paper. Aqueous solutions of 9% were obtained for each part of the plant. The methodology of extraction adopted is similar to that of other authors [20, 22, 23].

2. Germination Test and Root Length

The seeds were thoroughly washed with distilled water and surface sterilized with sodium hypochlorite (12%) for 2-3 minutes. Fifty seeds were placed in each sterilized Petri dish of 90 mm diameter and 15 mm height, and irrigated with 2 ml of water extracts (leaves, twigs, roots, seeds and pericarps at 28 °C. The control was treated with 2 ml distilled water.

Each treatment had four replicates. Analyzed seeds belong to six cultivated species: barley (*Hordeum vulgare*), alfalfa (*Medicago sativa*), jute (*Corchorus olitorius*), fenugreek (*Trigonella foenum-graecum*), lentil (*Lens culinaris*) and chickpea (*Cicer arietinum*). Every day, seed germination was investigated. Three days after germination, root length was measured. Data are given as percent of germination and radicle length.

3. Statistical Analysis

Analysis of variance (ANOVA) and the differences between the parameters were evaluated by Student-Newman-Keuls test. Differences were considered when $P \leq 0.05$. The used software is XLSAT, 2012.

RESULTS

1. Percentage of Germination

The effect of the aqueous extract of *J. curcas* varies according to plant organ. The extract of seeds had the lowest inhibitive effect for all species, the rate of germination with this extract being always over 70%. The extract of the pericarps had the highest inhibitive effect, with germination rates of 94.5 ± 3, 65.5 ± 1.9 and 22.5 ± 6.4 for *C. olitorius*, *H. vulgare* and *C. arietinum* respectively, and 0% for *M. sativa, T. foenum-graecum* and *L. culinaris*. The second inhibitive aqueous extract is that of leaves, the germination rate obtained with this extract is 94.5 ± 1, 47.5 ± 16.4, 43.5 ± 6.4, 30 ± 2.5, 30 ± 14.7 and 5.5 ± 3% for *C. olitorius, M. sativa, H. vulgare, C. arietinum, T. foenum-graecum* and *L. culinaris* respectively (Figure 1, 2, 3, 4, 5, 6). The aqueous extract of the twigs had the highest inhibitory effect on the germination of *L. culinaris*, its germination percentage is 37 ± 6.2% (Figure 6).

The aqueous extract of seeds had a limited effect on germination of all treated species, its elevated inhibitory effect was observed on the germination of *H. vulgare*; with a 70 ± 9.4% germination rate (Figure 3). Finally, the aqueous extract from roots had an elevated inhibitory effect on the germination of two species, *L. culinaris* and *C. arietinum*, resulting in germination rates of 23 ± 3 and 33.75 ± 2.5% respectively (Figure 4, 5).

The inhibitory effect of *J. curcas* aqueous extracts varies also with treated species; *C. olitorius* is the most tolerant species to the allelopathic effect caused by the aqueous extract of all organs of *J. curcas*. Highest germination percentage (94.5%) of this species was obtained with all extracts (Figure 2).

L. culinaris is the most sensitive species; its rate of germination is 97.5 ± 3, 23 ± 3.8, 37 ± 6.2, 5.5 ± 3 and 0% respectively for the control and the extracts of the roots, twigs, leaves and pericarps. But, this species (*L. culinaris*) is tolerant to the aqueous extract of seeds; we observed a germination rate of 91 ± 3.4% for this treatment (Figure 6).

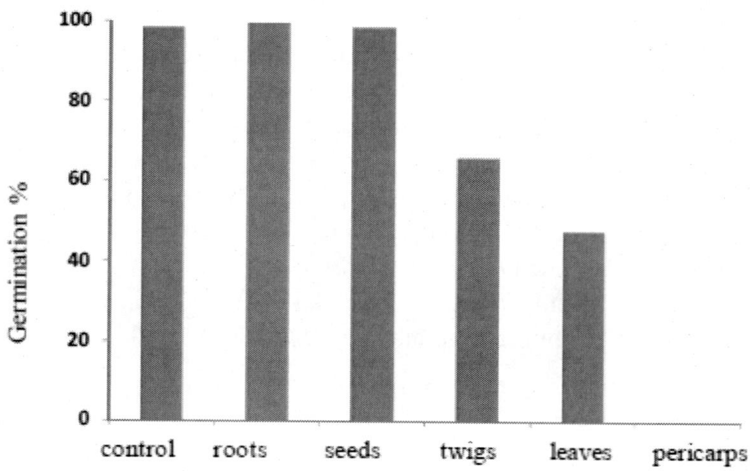

Figure 1. Effects of aqueous extract of *J. curcas* tissues on the germination of *M. sativa*.

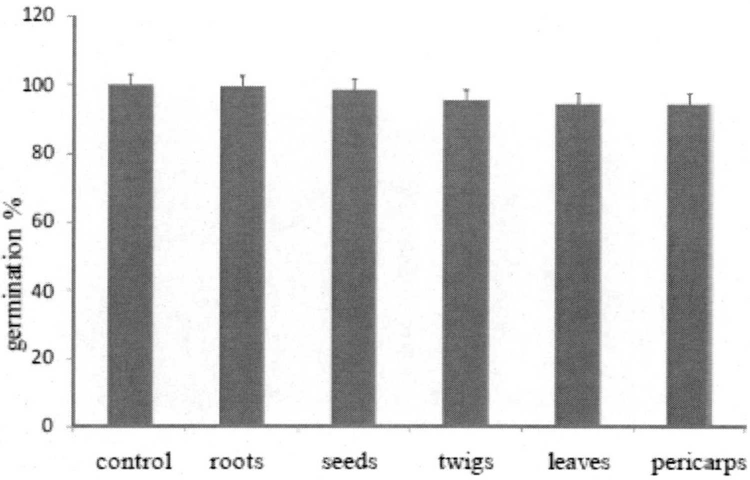

Figure 2. Effects of aqueous extract of *J. curcas* tissues on the germination of *C. olitorius.*

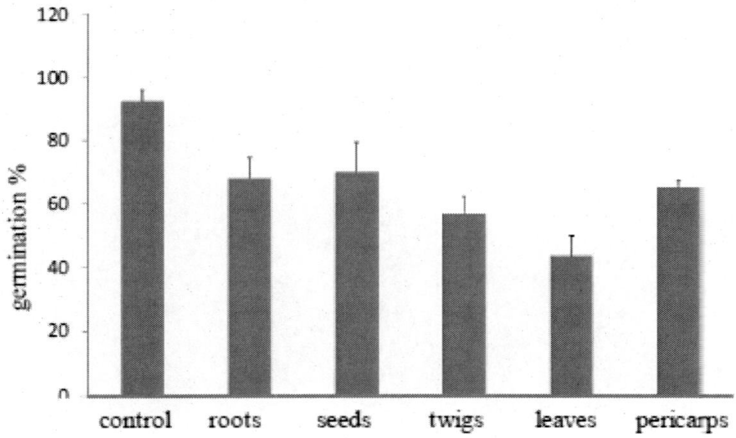

Figure 3. Effects of aqueous extract of *J. curcas* tissues on the germination of *H. vulgare.*

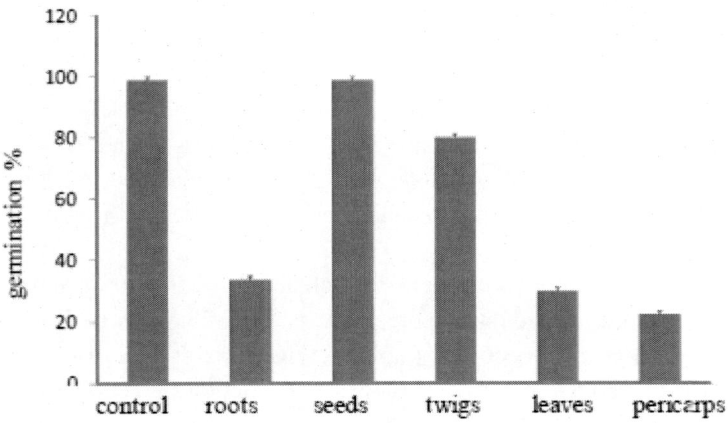

Figure 4. Effects of aqueous extract of *J. curcas* tissues on the germination of *C. arietinum.*

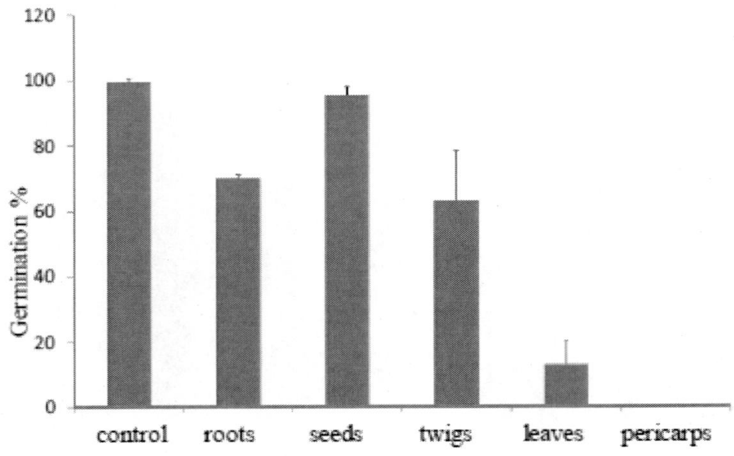

Figure 5. Effects of aqueous extract of *J. curcas* tissues on the germination of *T. foenum-graecum.*

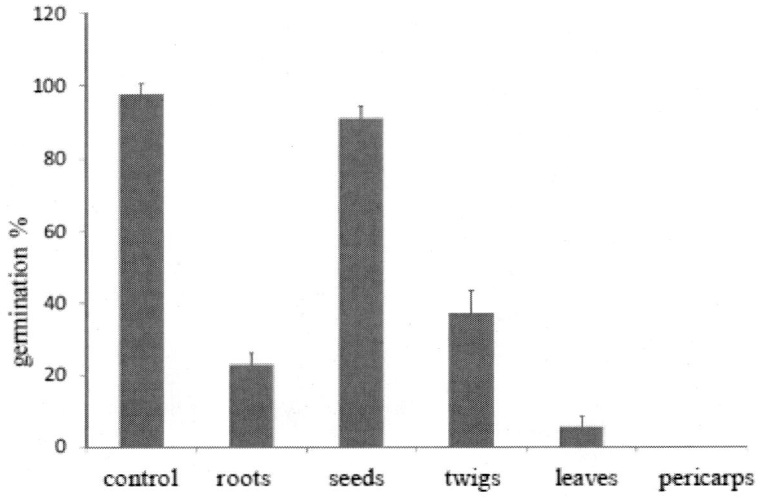

Figure 6. Effects of aqueous extract of *J. curcas* tissues on the germination of *L. culinaris.*

2. Root Length

All treated species showed a decrease in root length for all aqueous extracts (leaves, twigs, roots, seeds and pericarps). A single exception was observed in root development of barley with the root aqueous extract. But, the root length varies with the type of extract and the studied species.

An important reduction of root length was obtained for the six tested species with the pericarp aqueous extract; the reduction percentage is 87.1, 78.5 and 65.5% for *C. arietinum, C. olitorius* and *H. vulgare* respectively (table 1). The second level of root length inhibition was observed with aqueous extract from twigs; its reduction percentage is 48.3, 79.5, 27.3 and 87.1% for *M. sativa*, *C. arietinum, C. olitorius* and *H. vulgare* respectively (Table 1).

DISCUSSION

The aqueous extracts of *J. curcas* inhibit germination and root growth of six cultivated species studied in this work. Similar results were registered with other species; high inhibitory effect of aqueous extracts (8-10%) of root and leaf of *J. curcas* was observed on the germination and the radical length of *Phaseolus vulgaris, Zea mays, Lycopersicon lycopersicum* and *Hibiscus esculentus* [20].

Also, leaf extracts of different accessions of *J. curcas* inhibited the germination of *Triticum aestivum, Brassica juncea, Sesamum orientale* and *Vigna mungo* [17, 24]. The inhibition percent on the root growth of cress, lettuce, alfalfa, Italian ryegrass, barnyard grass, and crabgrass by *J curcas* extracts was 32, 61, 12, 21, 18, and 59, respectively, at concentration 30-mg dry weight equivalent extract/mL [19]. In addition, inhibitory effect of *Jatropha* rhizosphere soil on the germination of *Zea mays, Vigna radiata* and *Brassica campestris* was reported by Uttam et al. [25]. But, this allelopathic effect in germination and growth is lower than that of other species such as *R. communis* (Euphorbiaceae) [19].

Highest inhibitory effect in germination rate was observed with aqueous extract of *J. curcas* pericarp for *T. foenum-graecum, L. culinaris, M. sativa* and *C. arietinum* (0, 0, 2, 22.5% respectively). The presence of phenolic compounds could be the cause of reduction in germination and growth of seedlings in the test crops [20].

Moreover, *J. curcas* fruit is rich in phenolics [1], and the phenolic allelochemicals could inhibit cell division and alter the ultrastructure of the cells [26]. Pericarp is mechanically removed from the fruit in the first step during oil extraction and about one ton can be obtained from one hectare.

This material can be used in fertilization or as a source of energy [27]. In the case of pericarp used in fertilization, the choice of crop species is important because pericarp has a high allelopathic effect. Also, *J. curcas* leaf is rich in phenols, the leaf extract contained 0.463 mg/ml of phenols [28] and its aqueous extract had a high inhibition of the germination and root growth of studied species.

Corchorus olitorius is the most tolerant species to the inhibitive effect of aqueous extracts of all organs of *J. curcas*, the tolerance was observed with the extract of other species. Indeed, the germination rate of this species (*C. olitorius*) treated with aqueous extract (9%) of the root, the twig and the leaf of *Eucalyptus occidentalis, Acacia ampliceps* and *Prosopis juliflora* is higher than 90% [22].

On the other hand, *Ludwigia decurrens* and *L. adscendens* exudates had no inhibitory effect on the germination percentage of *C. olitorious* [1], but, the exudates from the two *Ludwigia* spp. induced mortality of 15 day old seedlings and a significant decrease in seedling elongation of *C. olitorious* [29], though, important inhibition of aqueous extract of *Echinochloa colona* on the germination of *C. olitorius* (63%), and a low extension of roots were registered [3]. In addition, the application of leaf leachates of *J. curcas* in the soil inhibited the shoot and root length and increased the relative membrane permeability and proline content in the roots of marigold (*Tagetes erecta*) seedlings [30].

For these authors, *J. curcas* residues incorporated into soil were more phytotoxic to the root than shoot growth of marigold seedlings. Also, in the field, the distance between *J. curcas* and other species is important; the effect of *J. curcas* hedgerow distance on growth and yield of maize (*Zea mays*) in the first year is no significant.

Table 1. Average root length of each species treated with aqueous extract of different organ of *J. curcas*

Species	*M. sativa*	*C. olitorius*	*H. vulgare*	*C. arietinum*	*T. foenum-graecum*	*L. culinaris*
Control	57.28 ± 9.56	63.06 ± 8.01	74.67 ± 11.01	93.22 28.09	72.22 ± 12.35	37.13 ± 5.79
Root	20.78 ± 3.19	23.76 ± 4.4	20.64 ± 4.4	31.81 ± 7.7	28.78 ± 5.25	25 ± 6.5
Seed	5.71 ± 2.2	18.06 ± 3.9	18.48 ± 4.6	23.98 ± 5.85	11.47 ± 1.95	11.11 ± 3.4
Twig	10.1 ± 1.9	18.66 ± 2.1	26.97 ± 7	28.77 ± 11.2	16.14 ± 4.92	20.75 ± 4.3
Leaf	7.88 ± 1.6	17.91 ± 1.9	12.8 ± 3.8	19.87 ± 2.9	8.33 ± 1.51	8 ± 0.85
Pericarp	0	2 ± 0.7	5.33 ± 1.9	3 ± 1.41	0	0

In the second year, significant differences were observed for some parameters; the closest distance from hedgerow (1 m) gave lowest plant height, diameter and stover weight, grain weight, weight of cob, weight of seed/cob [31]. However, for other species (*Brassica oleracea*), the diameter, stem length, leaf length, leaf area, fresh and dry mass of leaves and roots were influenced positively by the high concentrations of crude aqueous extract [32].

Jatropha plantation allowed healthy growth of other natural vegetations around it i.e. *Cynodon dactylon, Cannabis sativus, Solanum* sp., *Lantana camara, Cyperus rotundus, Cenchrus setigerus, Cassia tora, Chenopodium murale, Cassia tora, Cenchrus ciliaris, Parthenium hysterophorus, Ageratum conyzoides,* etc. [33].

CONCLUSION

High concentration (9%) of aqueous extract of *J. curcas* parts had a strong inhibitory effect on both germination rate and radicle length of studied species. The effect varies between species and depending on the process considered. On germination, *Corchorus olitorius* is the most tolerant species and *L. culinaris* is the most sensitive. For root length, *H. vulgare* is the most tolerant to *J. curcas* aqueous extract. Finally, the pericarp extracts had the highest allelopathic effect for the two physiological parameters (germination percentage and radicle length).

REFERENCES

[1] Sanderson, K., Bariccatti, R. A., Primieri, C., Viana, O. H., Viecelli, C. A., Bleil Junior, H. G. Allelopathic influence of the aqueous extract of jatropha on lettuce (*Lactuca sativa* var. Grand Rapids) germination and development. *Journal of Food, Agriculture and Environment*, 2013, 11(1), 641-643.

[2] Siddiqui, S., Ruchi Yadav, R., Yadav, K., Wani, F. A., Meghvansi, M. K., Sharma, S., Jabeen, F. Allelopathic Potentialities of Different Concentration of Aqueous Leaf

Extracts of Some Arable Trees on Germination and Radicle Growth of *Cicer arietinum* Var. – C-235. *Global Journal of Molecular Sciences*, 2009, 4(2), 91-95.

[3] Gomaa, N. H., AbdElgawad, H. R. Phytotoxic effects of *Echinochloa colona* (L.) Link. (Poaceae) extracts on the germination and seedling growth of weeds. *Spanish Journal of Agricultural Research*, 2012, 10(2), 492-501.

[4] Weir, T. L., Park, S., Vivanco, J. M. Biochemical and physiological mechanisms mediated by allelochemicals. *Curr. Opin. Plant Biol.*, 2004, 7, 472-479

[5] Kamal, S., Manmohan, S., Birendra, S. A review on chemical and microbiological applications of J*atropha curcas*. *International Research Journal of Pharmacy*, 2011, 2 (4), 61-66.

[6] Yue, G. H., Lo, L. Ch., Sun, F., Cao, S. Y., Yi, Ch. X., Hong, X., Sun, W. B. No Variation at 29 Microsatellites in the Genome of *Jatropha curcas*. *Journal of Genomics*, 2014, 2, 59-63.

[7] Behera, S. K., Srivastava, P., Tripathi, R., Singh, J. P., Singh, N. Evaluation of plant performance of *Jatropha curcas* L. under different agro-practices for optimizing biomass – A case study. *Biomass and Bioenergy*, 2010, 34, 30-41.

[8] Brittaine, R., Lutaladio, N. B. Jatropha: A Smallholder Bioenergy Crop: The Potential for Pro-Poor Development. *Integrated Crop Management,* Vol. 8. FAO Rome (Italy); 2010.

[9] Soliman, W. M., He, X. R. The Potentials of Jatropha Plantations in Egypt: A Review, *Modern Economy,* 2015, 6, 190-200, http://dx.doi.org/10.4236/me.2015.62016.

[10] Sano, R., Ara, T., Akimoto, N., Sakurai, N., et al. Dynamic metabolic changes during fruit maturation in *Jatropha curcas* L. *Plant Biotechnology*, 2012, 29, 175-178.

[11] Sundari, J., Selvaraj, R. *In vitro* evaluation of antibacterial and antifungal activity of *Jatropha curcas* L. *Rom. J. Biol. – Plant Biol.*, 2012, 57(1), 37-42.

[12] Rachana, S., Tarun, A., Rinki, R., Neha, A., Meghna, R. Comparative Analysis of Antibacterial Activity of *Jatropha curcas* Fruit Parts. *Journal of Pharmaceutical and Biomedical Sciences*, 2012, 15(15) file:///C:/Users/Master.Info/Downloads/Documents/Tarun%20Agarwal%20%20et.al.%281%29.pdf.

[13] Nwauzoma, A. B., Dappa, M. S. Ethnobotanical studies of Port Harcourt Metropolis, Nigeria. *American Journal of Research Communication*, 2013. www.usa-journals.com.

[14] Rodrigues, L. J. M., Nunes, M. D., Paes, S. A., Torres, D. P., Kasuya, M. C. M. Bio-detoxification of *Jatropha curcas* seed cake by *Pleurotus ostreatus*. *African Journal of Microbiology Research*, 2014, 8(11), 1148-1156. http://www.academicjournals.org/AJMR.

[15] Dubey, G., Bharati, K., Ahirwar, U., Mishra, R., Tiwari, S., Mohanty, S. R. Prospect of bioactive molecules from *Jatropha curcas* to improve soil and microbial quality for sustainable agricultura. *African Journal of Microbiology Research,* 2015, 9(3), 140-147.

[16] Ma, Y., Chun, J., Wang, S., Fang Chen, F. Allelopathic potential of *Jatropha curcas*. *African Journal of Biotechnology*, 2011, 10(56), 11932-11942.

[17] Venkatesh, A., Tapasya, S., Kumar, R. V., Gurunathan, N. Allelopathic effect of different accessions of *Jatropha curcas* on firld crops. *Range Mgmt. and Agroforestry*, 2011, 32(1), 40-44.

[18] Cremonez, P. A., Feiden, A., Santos, R. F., Bassegio, D., Rossi, E., Nadaleti, W. C., Antonelli, J., Tomassoni, F. Allelopathic influence of the aqueous extract of *Jatropha*

curcas L. leaves on wild *Cichorium intybus*. *African Journal of Agricultural research*, 2013, 8(49), 6575-6578.

[19] Mominul Islam, A. K. M., Kato-Noguch, H. Allelopathic prospective of Ricinus communis and *Jatropha curcas* for bio-control of weeds, *Acta Agriculturae Scandinavica, Section B - Soil and Plant Science*, 2014, 63(8), 731-739, DOI:10.1080/09064710.2013.865073.

[20] Aburge, S., Quashie-Sam, S. J. Q. Evaluating the allelopathic effect of *Jatropha curcas* aqueous extract on germination, radical, and plumule length of crops. *Int. J. Agric. Biol.*, 2010, 12, 769-772.

[21] Rejila, S., Vijayakumar, N., Jayakumar, M. Chromatographic determination of allelochemicals (phenolics acids in *Jatropha curcas* by HPTLC). *Asian Journal of Plant Science and Research*, 2012, 2(2), 123-128.

[22] Saadaoui, E., Ghazel, N., Ben Romdhane, Ch., Tlili, N., Khaldi, A. Allelopathic Effects of aqueous extracts of *Eucalyptus occidentalis* Endl., *Acacia ampliceps* Maslin and *Prosopis juliflora* L. on the germination of three species cultivated in the South of Tunisia. *Tunisian Journal of Plant Protection*, 2014, 9(1), 11-16.

[23] Nadaletti, W. C., Bariccatti, R. A., Santos, R. F., et al. Influence of leaf extract of *Jatropha curcas* on initial growth of cauliflower (*Brassica oleracea* var. *botrytis*). *Afr. J. Agric. Res.*, 2014, 9(4), 521-526.

[24] Reichel, T., Barazetti, J. F., Stefanello, S., Paulert, R., Zonetti, P. C. Allelopathy of leaf extracts of jatropha (*Jatropha curcas* L.) in the initial development of wheat (*Triticum aestivum* L.). *IDESIA* (Chile), 2013, 31(1), 45-52.

[25] Uttam, D. M., Khare, N., Karanjalker, G. R. Assessment of phytotoxic proclivities of *Jatropha curcas* L. on germination and seedling establishment of some field crops. *International Journal of Agriculture, Environment and Biotechnology*, 2014, 6(1), 113-120.

[26] Gomaa, N. H., Hassan, M. O., Fahmy, G. M., González, L., Hammouda, O., Atteya, A. M. Allelopathic effects of *Sonchus oleraceus* L. on the germination and seedling growth of crop and weed species. *Acta Bot. Bras.*, 2014, 28(3): 408-416.

[27] Jingura, R. M., Musademba, D., Matengaifa, R. An evaluation of utility of *Jatropha curcas* L. as a source of multiple energy carriers. *International Journal of Engineering, Science and Technology*, 2010, 2(7), 115-122.

[28] Oloyede, O. B., Salau, A. K., Akeusola, R., Ganiyu, O. T., Azeez, L., Ogunbode, S. M. Phytochemical Content, Radical Scavenging and Antibacterial Properties of Aqueous Extract of *Jatropha curcas* Linn Leaves. *Fountain Journal of Natural and Applied Sciences*, 2012, 1(1), 41-4.

[29] Sakpere, A. M., Oziegbe, M., Bilesanmi, I. A. Allelopathic Effects of *Ludwigia decurrens* and *L. adscendens* subsp. *diffusa* on Germination, Seedling Growth and Yield of *Corchorus olitorious* L. *Notulae Scientia Biologicae*, 2010, 2(2), 75-80.

[30] Wang, J. C., Wu, Y., Wang, Q., Peng, Y. L., Pan, Y., Luo, P., Wu, N. Allelopathic effects of *Jatropha curcas* on marigold (*Tagetes erecta*). *Allelopathy Journal*, 2009, 24 (1), 123-130.

[31] Abugre, S., Oti-Boateng, C., Adu-Gyamfi, A. Effect of distance from *Jatropha curcas* hedgerow on growth and yield of *Zea mays*. *Agric. Biol. J. N. Am.*, 2011, 2(3), 471-475.

[32] Maharjan, S., Shrestha, B. B., Jha, P. K. Allelopathic effects of aqueous extract of leaves of *Parthenium hysterophorus* l. on seed germination and seedling growth of some cultivated and wild herbaceous species. *Scientific World*, 2007, 5(5), 33-39.
[33] Sahoo, N. K., Kumar, A., Sharma, S., Naik, S. N. Interaction of *Jatropha Curcas* Plantation with Ecosystem. *Proceedings of International Conference on Energy and Environment*, 2009, 19-21.

In: Jatropha Curcas
Editor: Gregory Medina

ISBN: 978-1-63483-089-8

Chapter 8

SIMPLE AND EFFECTIVE METHOD FOR REMOVAL OF PHORBOL ESTERS FROM *JATROPHA CURCAS* SEED OIL BY SILICA GEL ADSORPTION

Wimonrat Tongpoothorn[1], Warunthip Chatjutamanee[1], Panuwat Supprung[1] and Chalerm Ruangviriyachai[2*]

[1]Department of Chemistry, Faculty of Engineering,
Rajamangala University of Technology Issan, Khon Kaen Campus,
Khon Kaen, Thailand

[2]Department of Chemistry, and Center of Excellence for Innovation
in Chemistry, Faculty of Science, Khon Kaen University, Khon Kaen, Thailand

ABSTRACT

The removal of phorbol esters (PEs) from *Jatropha curcas* seed oil by adsorption onto silica gel was investigated. The average PE content in the oil, as determined by HPLC, was 1.68 mg TPA/g oil. The optimized conditions for PE adsorption by silica gel were studied. The results showed that the highest PEs removal (77.5%) and PE adsorption (104.2%) were obtained when 0.3 g of silica gel/5 g of seed oil was used at ambient temperature (30 °C) for 60 min of adsorption time and with an agitation rate of 250 rpm. The characteristics of the silica gel, as determined by FT-IR, indicated that PEs were adsorbed onto the surface of the silica gel. A comparison of the physico-chemical properties of the raw oil and the oil obtained after adsorption demonstrates that the adsorption by silica gel let to receive lightens the oil color without changing its fatty acid composition or introducing contamination from other compounds. Additionally, the results showed that silica gel can be used two additional times in subsequent PE adsorptions without a significant loss in reusability. In conclusion, adsorption by silica gel can remove PEs from *J. curcas* seed oil. This method offers simplicity of design, operation and scale-up, avoids using toxic solvents and minimizes the degradation of PEs.

---

* Corresponding author's email: chal_ru@kku.ac.th.

Keywords: Phorbol ester, *Jatropha curcas* seed oil, silica gel, adsorption

1. INTRODUCTION

Jatropha curcas (*J. curcas*) has been investigated primarily as a source of oil. The seed cake that remains after oil extraction is an excellent source of plant nutrients. However, the presence of high levels of toxic substances and anti-nutrients limits the further applications of *J. curcas* oil and meal. The main toxins present in *J. curcas* are phorbol esters (PEs). The term 'phorbol esters' is used to describe a naturally occurring family of compounds widely distributed in plant species of the families *Euphorbiaceae* and *Thymelaeceae* (Haas and Mittelbach, 2000). Although one specific phorbol ester, phorbol-12-myristate-13-acetate, has been identified as the major toxic compound in *J. curcas* (Makkar et al., 1998), the PE content is dependent on the soil and climatic conditions (Martínez-Herrera et al., 2006). PEs are co-carcinogens, purgative activities and skin irritants. In human, accidental poisoning by physic nut seed has been reported to elicit giddiness, vomiting and diarrhea. PEs, even at low concentrations, show toxicological effects in animal feed diets as report by Goel et al. (2007). *J. curcas* oil and its extract can exhibit insecticidal effects, pesticidal effects, and antimycobacterial activity and also offer the potential for snail-killing effects (Chumkaew et al., 2003, Makkar and Becker, 2009, Diwani et al., 2009). Many researchers have studied the planting and testing of *J. curcas* and biodiesel fuel derived from its oil, as well as the removal of toxic substances from the oil and meal. The problem is the detoxification step that has to be developed. PEs are highly unstable and auto-oxidize under normal storage conditions; however, they appear to be stabilized by some other components of *J. curcas* oil. PEs appear to be highly stabilized, so neither the temperatures nor the pressures common in compression processes can destroy by these compounds (Makkar et al., 1997). Until now, no methods for the complete removal of PEs from *J. curcas* have been reported. In previous efforts, researchers have attempted to eliminate phorbol esters from the oil and seed using physical, chemical and even biological treatments (Haas and Mittelbach, 2000, Martínez-Herrera et al., 2006, Siddhuraju et al., 2002, Aregheore et al., 2003, Ahmed and Salimon, 2009, Donlaporn and Worapot, 2011, Belewu and Sam, 2010, Joshi et al., 2011). However, these methods involved considerable technical and economic efforts and require immense solvent consumption. The optimized conditions for such treatments have not been defined. In some cases, the oil was contaminated with some chemicals or microorganisms that required further refining for their removal, which resulted in intense energy costs. These issues provided the basis for an investigation in the present work to eliminate PEs from the oil using adsorption techniques that are simple, rapid and environmentally friendly. The extraction of PEs from the oil was conducted to turn PEs into some value-added products, such as biopesticides for organic as well as conventional agriculture. The fatty acid composition of *J. curcas* seed oil is close to that of olive oil (Makkar and Becker, 2009). After the elimination of PEs in *J. curcas* seed oil, it can has broader applications.

The aim of this work was to investigate the removal of PEs from *J. curcas* seed oil through silica gel adsorption. The effects of the amount of silica gel, the adsorption temperature, the adsorption time and the agitation rate on PE removal and PE adsorption were investigated. The characteristics of the oil and silica gel, including the reusability of the silica gel, were also studied.

2. Materials and Methods

2.1. Materials

J. curcas seed was supplied by the Office of Agricultural Research and Development, Region 3, Khon Kaen Province and by the Nakon Ratchasima Field Crops Research Center, Nakon Ratchasima Province, Thailand. The seed was pressed to obtain the oil using an oil expeller from the Agricultural Engineering Research Centre Khon Kaen, Khon Kaen Province, Thailand. The extraction solvents (methanol, acetonitrile and hexane) were of analytical grade and purchased from Carlo Erba (Italy), whereas HPLC-grade solvents were obtained from Lab Scan (Ireland). Standard phorbol ester (tetradecanoyl phorbol-13-acetate, TPA) used as an external standard, was purchased from Sigma-Aldrich (Switzerland). The adsorbent (chromatography silica gel with a particle diameter of 0.06-0.2 mm) used in this study was obtained from Carlo Erba (Italy). A standard mixture of fatty acid methyl esters (FAMEs) was purchased from Sigma-Aldrich (UK).

2.2. Optimized Conditions for PE Adsorption

Amount of Silica Gel

Each 5.00 g of *J. curcas* seed oil was mixed with silica gel, and the weight of the silica gel was 0.1, 0.2, 0.3, 0.4 and 0.5 g. The mixtures were stirred using a shaker operated at 150 rpm for 60 min at ambient temperature (30°C). Then, 5.0 mL of hexane was added to each mixture, and the mixtures were subsequently filtered through a filter paper to separate the oil from the silica gel. Each portion was then extracted and analyzed further for PE content.

Adsorption Temperature

The optimum conditions obtained from the results of silica-gel experiments described in mention above were used in subsequent studies. Each 5.00 g of the oil was mixed with the appropriate amount of silica gel. The mixtures were stirred using a shaker operated at 150 rpm for 60 min with varying adsorption temperatures: ambient temperature (30°C), 45°C, 60°C and 75°C. Then, 5.0 mL of hexane was added to each mixture, and the mixture was filtered through a Whatman filter paper (UK) to separate the oil from the silica gel. Each portion was extracted and analyzed for PE content.

Adsorption Time

The optimum quantity of silica gel and the optimum adsorption temperature were used in subsequent studies. Each 5.00 g of the oil was mixed with the appropriate amount of silica gel. The mixtures were stirred using a shaker operated at 150 rpm. The adsorption process was performed using different contact times (15, 30, 60, 90 and 120 min) at ambient temperature (30°C). 5.0 mL of hexane was then added to the mixture, and the mixture was filtered through a Whatman filter paper to separate the oil from the silica gel. Each portion obtained was subjected to the extraction and also analysis of PEs.

Agitation Rate

Under the optimized conditions, each 5.00 g of the oil was mixed with the appropriate amount of silica gel. The mixtures were stirred with a shaker operated at 50, 100, 150, 200 and 250 rpm at the optimum temperature under the optimum adsorption time. 5.0 mL of hexane was then added to each mixture, and the mixtures were filtered through a Whatman filter paper to separate the oil from silica gel. Each portion was further extracted and analyzed for PEs.

2.3. Extraction of PEs

PEs in the oil and PEs adsorbed on silica gel were extracted using a slightly modified version of the method of Rakshit et al. (2008). The oil or silica gel was extracted for four times using mechanical extraction as follows: the oil or silica gel was mixed with methanol (1:2 w/v) and stirred (300 rpm) at 60°C for 5 min. The methanol phase was separated from the oil and silica gel by gravity. The remaining oil and silica gel were re-extracted for three times under the same conditions. The methanol phases were combined and frozen for 12 h to provide a partial separation of the oil. The combined methanol extracts were then filtered, and the filtrate was evaporated to remove the solvent using a vacuum evaporator. The residue was re-dissolved with a portion of acetonitrile (3.00 mL) and extracted with hexane (2.00 mL) so that the remaining oil could be removed using a separatory funnel. The acetonitrile phase was then evaporated and re-dissolved with HPLC-grade methanol to obtain the PE-rich fraction for further analysis.

2.4. Analysis of PEs Using HPLC

PEs in each extract from the oil and silica gel were analyzed and quantified using a Shimadzu HPLC system, which consisted of a Shimadzu LC-20AD pump, a DGU-20A5 degasser, and a Shimadzu SPD-M20A photodiode array detector (Japan). All of solutions used in this research were prepared with deionized water (Simplicity Water Purification System, model Simplicity 185, Millipore Corporation, USA). A reverse-phase Hypersil ODS, C_{18} (250 × 4.6 mm i.d., 5.0 μm) column was used in this study. The column was protected with a guard column that contained the same materials. The separation was performed according to the procedure described by Haas and Mittelbach (2000). The mobile phase used was acetonitrile and water (90:10 v/v) with flow rate 1.0 ml/min. An isocratic elution was used. The mobile phase and sample were filtered through a 0.45 μm filter membrane and degassed before analysis. The phorbol ester peaks were detected at 280 nm, and the elution of the phorbol ester peaks was determined to occur between 9 and 15 min. Phorbol-12-myristate-13-acetate (TPA) was used as an external standard. The area of the phorbol ester peaks was summed and converted to a phorbol-12-myristate-13-acetate equivalent.

2.5. Characterization of the Oil and Silica Gel

Determination of Fatty Acid Compositions of the Oil Using a GC-FID

The fatty acid composition of *J. curcas* seed oil and the oil obtained after adsorption was determined. The method described by Tuberoso et al. (2007) was used to prepare the sample for the analysis of fatty acid methyl esters (FAMEs). This method was performed by pouring 0.5 g of oil and 250 μL of 2 M KOH in methanol into a 15 mL vial and mixing with a vibration mixer (Falc Instruments, Treviglio, BG, Italy) for 60 s. After 15 min of rest, 6 mL of *n*-hexane was poured into the vial, and the mixture was mixed with a vibration mixer for 10 min. The layers were allowed to separate, and the hexane fraction was injected into the GC for analysis. Fatty acids were determined with an Agilent trace GC chromatograph equipped with a methyl silicone HP-5 capillary column (30 m × 0.25 mm i.d., 0.25 μm film thickness). The injector temperature was 260°C. The oven temperature was programmed to increase from 70°C to 280°C at 1°C/min. The carrier gas was helium flowed at a rate of 1.0 mL/min (constant flow), and 10 μL of the extract was injected using the split mode. The temperature of the FID was set to 280°C. The fatty acid methyl esters were identified through a comparison of their retention times with the relative retention times of standard FAME peaks obtained under the same conditions. The experiments were done in triplicate. A standard mixture of FAMEs (C_8-C_{24}) was prepared in hexane. The percentage composition of the oil was calculated from the GC peak areas using the Xcalibur 3.0 software package.

Characterization of the Oils and Silica Gels Using a FT-IR

A few milligrams of silica gel and silica gel with adsorbed PE were ground and mixed with an appropriate amount of potassium bromide (KBr) powder, which had been heated in an oven at 110°C to remove humidity before use. The homogeneous powders were pressed into a transport disk. The disk sample was then scanned on a FT-IR spectrometer, Perkin Elmer, Spectrum One model (England) over the wavenumber range of 4,000 to 450 cm^{-1}. A few oil samples (before and after adsorption) were dropped onto NaCl cell plates and then scanned on a FT-IR spectrophotometer over the same wavenumber range.

Reusability of Silica Gel

The reusability of the silica gel was tested by adsorption five times under optimized conditions. After the first adsorption, 5.0 g of silica gel was regenerated by being mixed with 25.00 mL of hexane. The mixture was then shaken using an orbital shaker with an agitation rate of 250 rpm for 2 h at ambient temperature (30°C). After being shaken, the silica gel was separated from the hexane by filtration through no. 1 filter paper. The experiments were performed in triplicate.

The remaining silica gel was washed three additional times with 25.00 mL of methanol under the same conditions. After being regenerated, the silica gel was used for the next run. The PE adsorbed by the regenerated silica gel was extracted and analyzed using previously described methods. The characteristics of each silica gel obtained after regeneration were determined using a FT-IR. The experiments were performed according to the previously described process.

3. Results and Discussion

3.1. PE Content of *J. Curcas* Seed Oil

HPLC chromatograms of standard TPA and PEs from *J. curcus* seed oil are shown in Figure 1. The TPA peak appeared at approximately 19.7 min whereas PE peaks from the oil extract eluted in the range of 9-15 min, and this elution time was used to calculate the PE content. The PE content in the oil was calculated based on the peak area relative to that of the standard curve of TPA and was expressed as mg TPA equivalent/g of oil. In this study, the average PE content was 1.68 mg TPA/g oil.

The PE content is known to depend on the soil and climatic conditions (Martínez-Herrera et al., 2006). The oil from the non-toxic Mexican varieties of *J. curcas* seed has been reported to contain negligible or low amounts of PEs (0.27 mg/mL of oil), whereas the toxic varieties have been found to contain 2.49 mg/mL of oil. Various parts of the *J. curcas*, such as the kernel, leaves, stems, flowers, buds, roots, latex, bark and wood contain different weights of PEs over the range of 0.39-6.00 mg/g dry matter as phorbol-12-myristate-13-acetate (PMA) equivalent (Makkar and Becker, 2009). However, the PEs are located mainly in the kernel portion of the seed, and the concentration of PEs varies with the genotype in ranging from 0.80-3.30 mg/g kernel (Makkar et al., 1998).

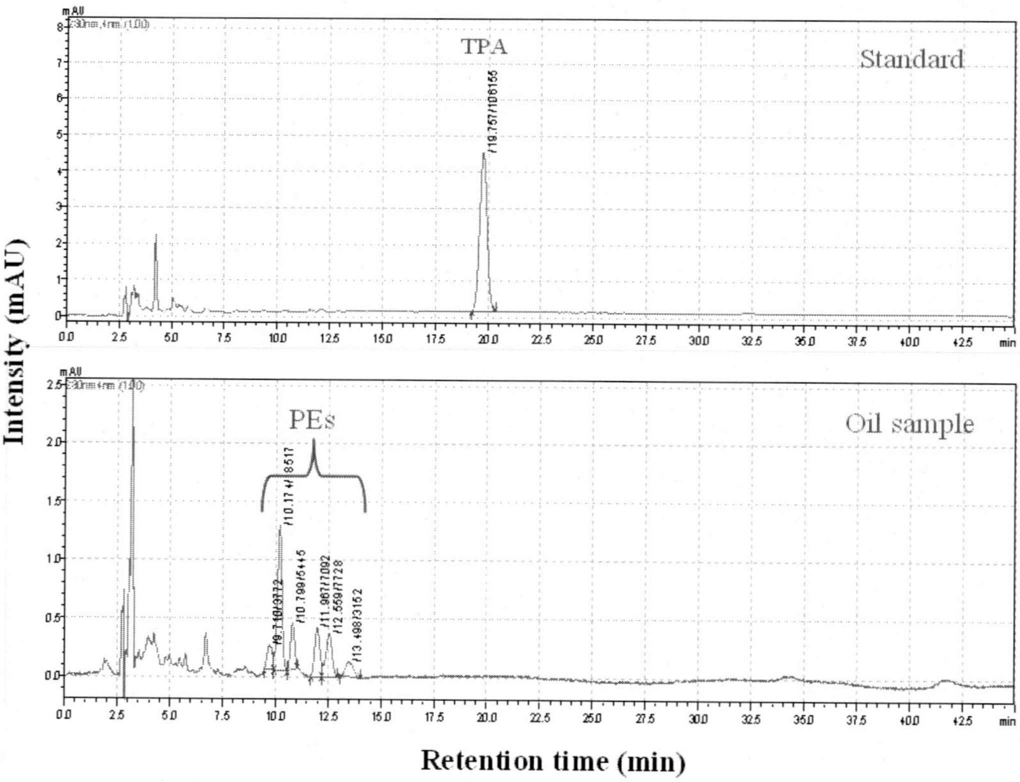

Figure 1. HPLC chromatograms of standard TPA and phorbol esters from *J. curcas* seed oil.

3.2. Optimized Conditions for PE Adsorption

Amount of Adsorbent

The amount of adsorbent is particularly important because it determines the extent of PE removal and may also be used to predict the cost of silica gel per unit of the oil to be tested. In these experiments, the determination of PEs in both the oil and the silica gel was performed; the results are shown in Figure 2. The percentage of PEs removed and adsorbed increased with increasing amounts of silica gel up to 0.3 g. An increase in the number of adsorbent surfaces and the availability of more adsorption sites can explain the increased percentage of PE removal. When the amount of silica gel used was greater than 0.4 g, the percentage of PE removal and adsorption decreased, possibly because overlap or aggregation of the adsorption sites resulted in a decrease in the total adsorbent surface area available to the PEs and in an increase in the diffusion path length (Crini et al., 2008). The time required to achieve equilibrium decreased with greater amounts of adsorbent. An amount of 0.3 g of silica gel, which resulted in the removal of 54.14% of the PEs and the adsorption of 116.41% of the PEs was therefore selected for further experiments.

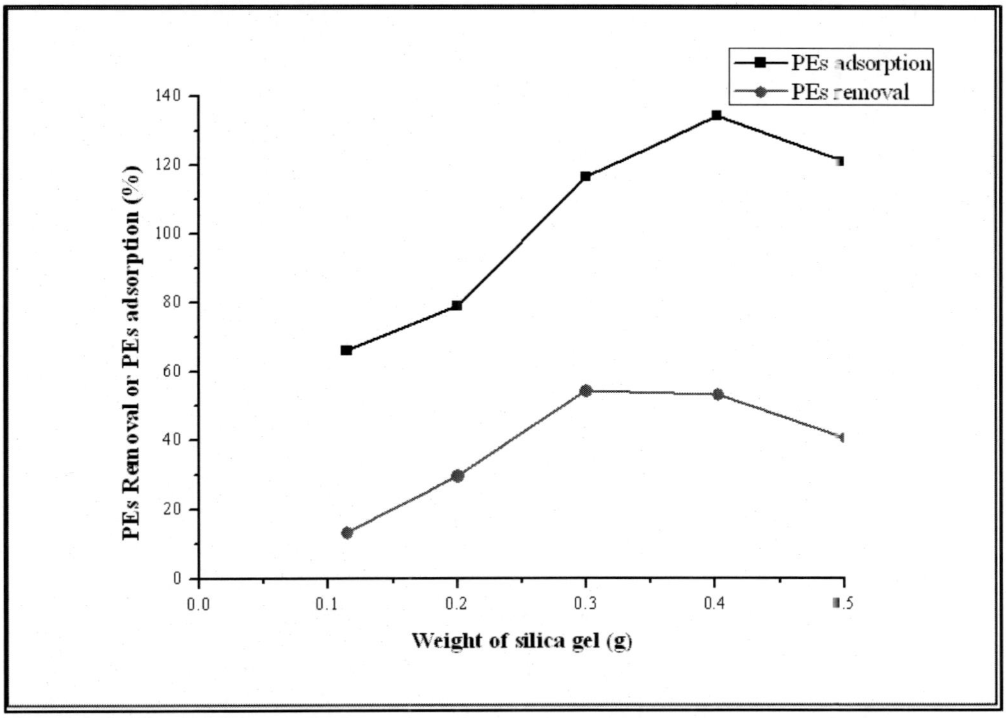

Figure 2. Percentage of PE removal and PE adsorption with different amounts of silica gel.

Adsorption Temperature

Adsorption typically increases with temperature because high temperature can produce a faster rate of diffusion of adsorbed molecules from the solution to the adsorbent. The percentages of PE removed from the oil and PEs adsorbed by silica gel at different adsorption temperatures in the range of 30-75°C are shown in Figure 3. The percentages of PEs removed and PEs adsorbed increased gradually with increasing adsorption temperature up to 45°C;

however, these values increased only slightly at higher temperature. These results can be explained as an increase in temperature resulting in an increase in the diffusivity of the adsorbed molecule and, consequently, in an increase in the adsorption rate of diffusion (Crini and Pierre-Marie, 2008). When the adsorption temperature at 75°C decreased in PEs removal and adsorption were observed. The temperature could also influence desorption and, consequently, the reversibility of the adsorption equilibrium. An increase in temperature was followed by a decrease in adsorption capacity, which suggests that adsorption is governed only by physical phenomena. The results of this experiment showed that the adsorption should be performed at high temperatures. However, the adsorption process is not usually conducted at such temperatures because operation at high temperatures increases operating costs. Therefore, the ambient temperature (30°C) resulted in 53.57% PE removal and 95.99% PE adsorption was selected for further experiments.

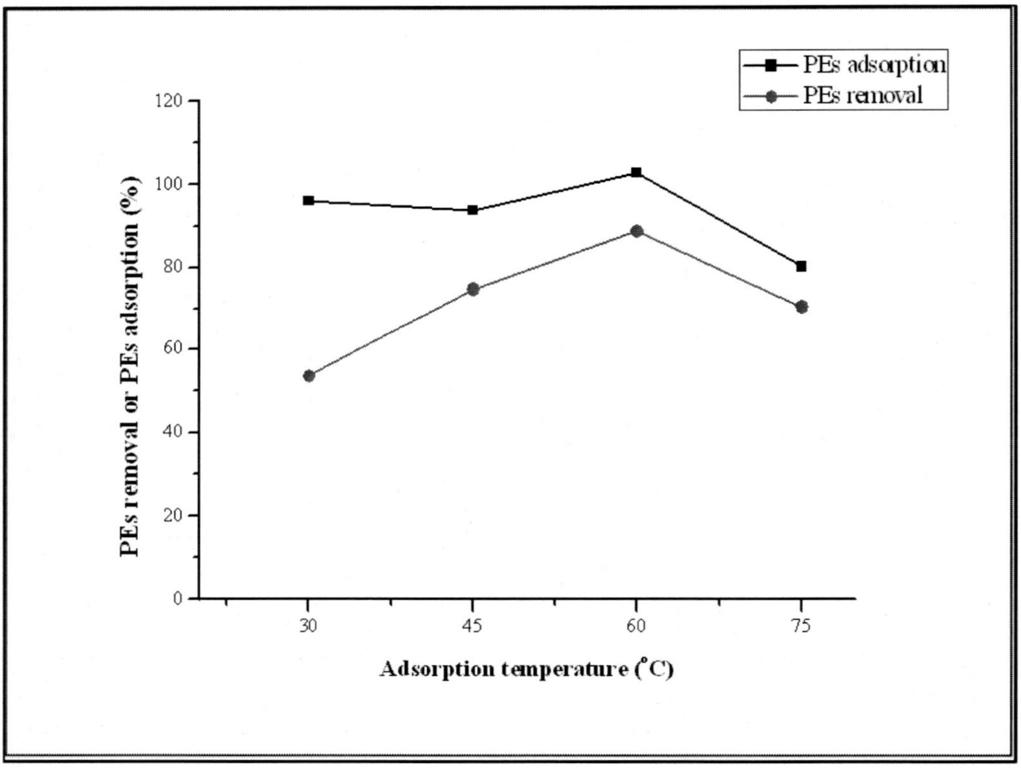

Figure 3. Percentage of PE removal and PE adsorption with different adsorption temperature.

Adsorption Time

Adsorption time is another important variable in adsorption processes. The adsorption capacity and the removal efficiency of PEs by silica gel typically increased with prolonged adsorption times. However, in practice, the adsorption time should be optimized with consideration of both the efficiency of desorption process and regeneration of the adsorbent. The percentages of PE removed and PE adsorbed are shown in Figure 4.

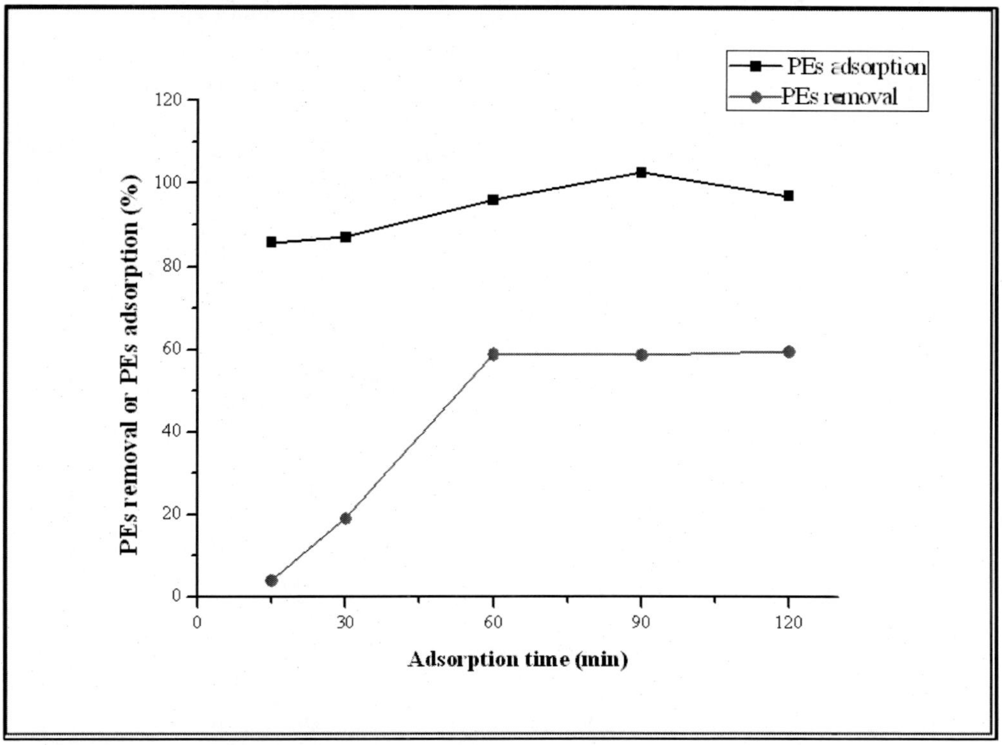

Figure 4. Percentage of PE removal and PE adsorption with different adsorption time.

The PE removal process occurred rapidly at the initial time. The percentage of PEs removed increased gradually with increasing adsorption time up to 60 min and subsequently reached a plateau. The percentage of PEs adsorbed as a function of the adsorption time slowly increased until the adsorption time reached to 90 min. However, a slight decreased in PE adsorption was observed when the adsorption time was at 120 min, most likely because the adsorption of PEs was rapid at the initial stages of the adsorption time and thereafter became slower near the equilibrium point. A large number of vacant surface sites are available for adsorption during the initial stage, and, after a lapse of time, the occupation of the remaining vacant surface sites was difficult due to repulsive forces between PE molecules adsorbed on the silica gel and PE molecules in the oil phase (Crini and Pierre-Marie, 2008). An adsorption time at 60 min produced 58.83% PE removal and 96.00% PE adsorption was therefore sufficient to reach adsorption equilibrium in further experiments.

Agitation Rate

Shaking is an important parameter in adsorption because it influences the distribution of the solute in the bulk solution and the formation of the external boundary film. These experiments were performed using an agitation rate that ranged from 50-250 rpm, and the results are shown in Figure 5. The shaking speed significantly affected the removal and adsorption of PEs. With an increase in shaking speed, the adsorption increased. When the shaking speed was slow, silica gel accumulated instead of spreading in the oil, and various active sites were buried under the layers above. PE removal appeared only by the layers above, in the layers below the boundary did not participate in the process because they have no contact with the PEs. The degree of agitation reduced the boundary-layer resistance and increases the mobility of the system. An increase in agitation through increasing the shaker speed results in a reduction of the external mass transfer effect. The same observation was made by previous work of Anwar et al. (2009). These results indicated that the agitation rate should be sufficient to assure that all the surface binding sites were readily available for PE uptake. An agitation rate at 250 rpm selected as the optimum rate resulted in 77.53% PE removal and 104.18% PE adsorption.

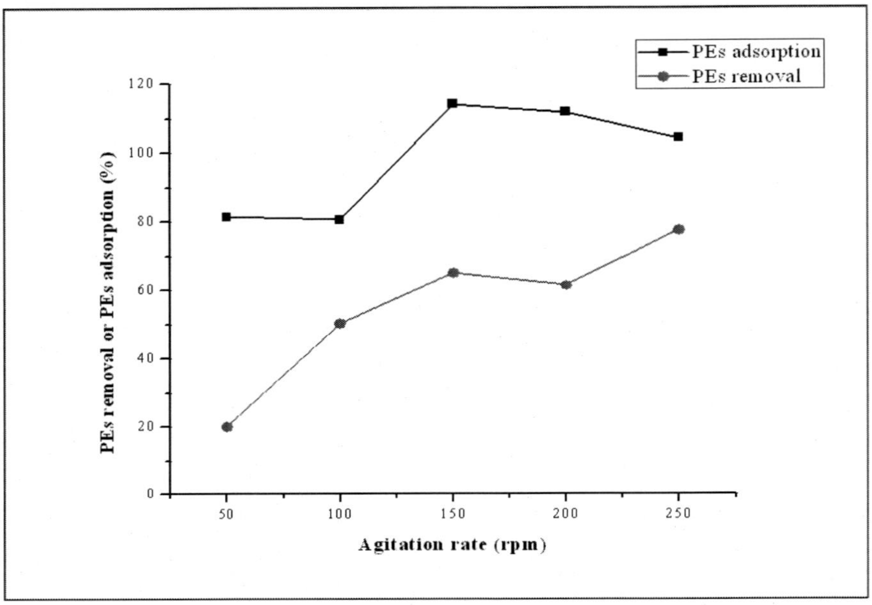

Figure 5. Percentage of PE removal and PE adsorption with different agitation rate.

3.3. Characteristics of the Oil by FT-IR

In this study, adsorption by silica gel changed the color of the oil from dark to bright-yellow. However, no change in the fatty acid composition was observed, as shown in Figure 6. This shows that the predominant fatty acid composition of both oils (the raw oil and the oil obtained after adsorption) in the order of highest relative percentage are palmitic acid (C16:0), stearic acid (C18:0), oleic acid (C18:1), linoleic acid (C18:2), linolenic acid (C18:3) and arachidic acid (C20:0). The fatty acid composition of the oil before and after adsorption was similar. These results are supported by FT-IR spectra, as shown in Figure 7. Typical FT-IR spectra of the raw oil contain structural or functional groups information on the fatty acids present in the oil, and the spectra are similar to the spectra of the oil obtained after adsorption. The spectra displayed three dominant regions related to fatty acids. The first region is the 3025-2800 cm^{-1} region in which methylene (CH_2) and methyl (CH_3) groups absorb. The second region is at 1740 cm^{-1} and is assigned to the carboxylic group (C = O), a region where the methyl ester and triglycerides absorb. The third region is 1500-900 cm^{-1}, also known as the fingerprint region.

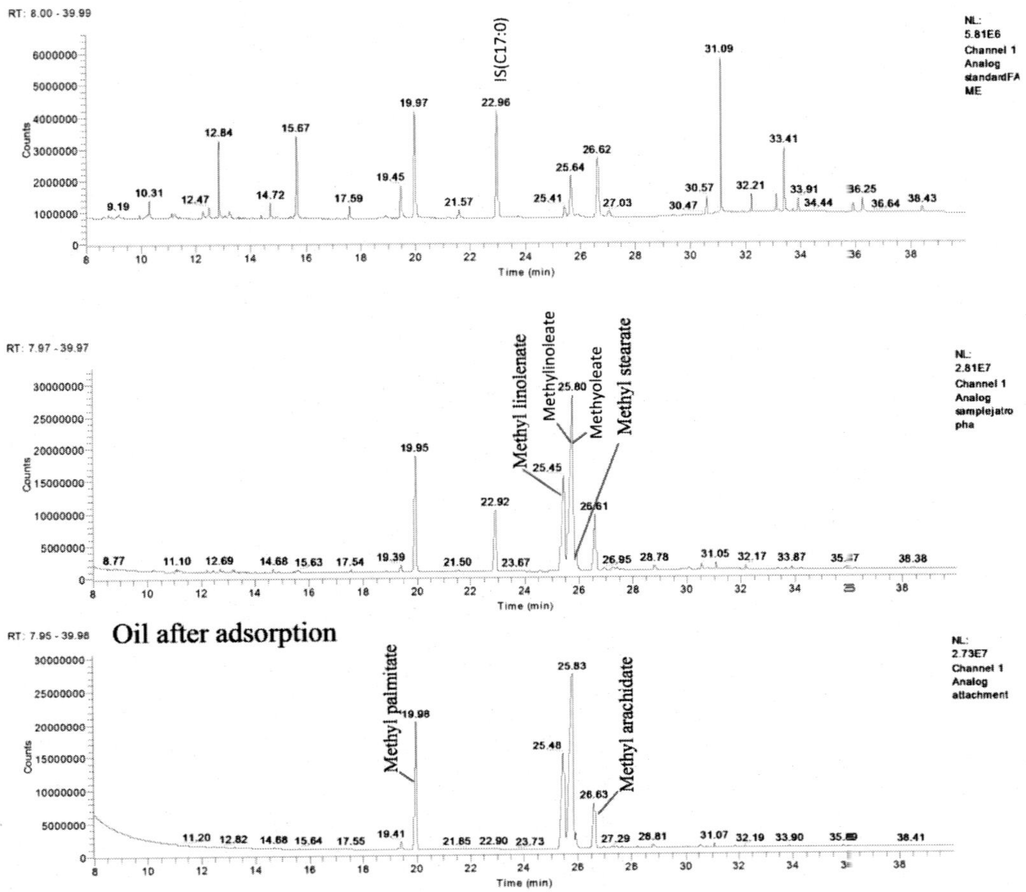

Figure 6. GC chromatogram of standard fatty acid methyl esters and the raw oil and the oil obtained after adsorption.

These results agreed with those works of Berghuis et al. (2010), who characterized fatty acids in soybean oil. A strong band at 1164 cm^{-1} was also observed to be flanked by weaker bands at 1237 and 1095 cm^{-1}, as observed in the spectrum of a triglyceride (Barua et al., 2008). Other bands observed in the fingerprint region are related to CH_3 and CH_2 symmetric and asymmetric deformations in the 1465-1350 cm^{-1} range and in the 1200-1000 cm^{-1} range. The band at 1021 cm^{-1} may involve stretching of the C-O bond. In addition, the band at 723 cm^{-1} is similar to the band observed in the spectrum of fatty acids and triglycerides and can be assigned to the CH_2 rocking mode. In conclusion, the FT-IR results show that the chemical properties of the raw oil and the oil obtained after adsorption on to silica gel are not different. The oil did not appear to be contaminated with other compounds after adsorption. Therefore, the oil adsorbed onto silica gel can be used for other applications with a little change in the chemical properties of the product, as seen in the studies of Makkar et al. (2009) and Devappa et al. (2010), who reported that biodiesel produced from the toxic variety of *J. curcas* seed oils presented quality as good as that produced from the non-toxic variety.

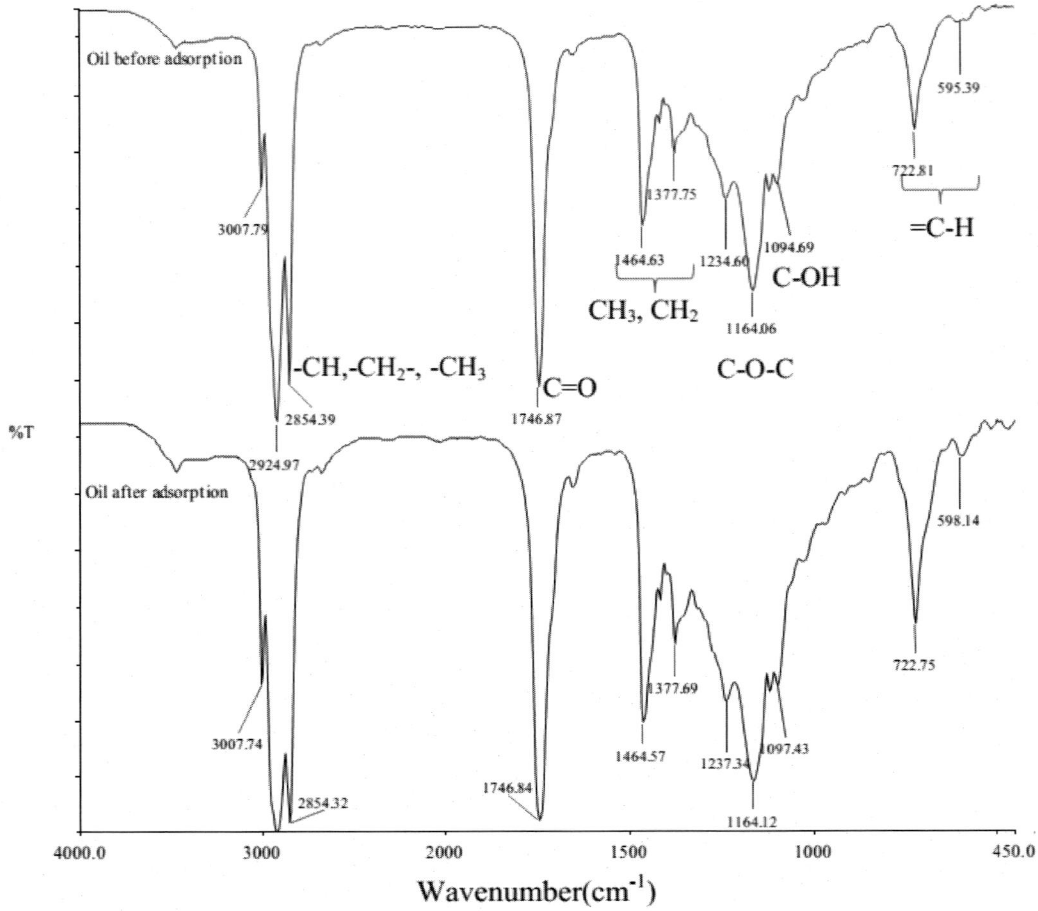

Figure 7. FT-IR Spectra of the raw oil and the oil obtained after adsorption.

3.4. Characteristics of Silica Gel Surface by FT-IR

The characteristics of the silica gel surfaces both before and after adsorption were studied to confirm that PE was adsorbed onto the surface of silica gel; the results are shown in Figure 8. The FT-IR spectra were used to characterize divided silica in the region of 3800-3200 cm^{-1}, which correspond to the OH stretching mode of silanols (Si-OH). The peak appeared at approximately 1094-1088 cm^{-1} and was assigned to asymmetric Si-O-Si stretching. The values were reported to be in the range of 970-967 cm^{-1} and 803-801 cm^{-1} for Si-OH bond stretching and OH bending (silanol), respectively. The presence of a peak at approximately 482-478 cm^{-1} was attributed Si-O bond rocking as previous works (Parida et al., 2006, Sasithorn et al., 2010).

The peak at 1638 cm^{-1} was attributed to the C = C bonds of unsaturated non conjugated aliphatic compounds. The presence of a peak at 1748 cm^{-1} in silica gel after adsorption, which is related to the C = O of the ester, was observed. Esters not only show their carbonyl C = O stretch at 1750-1735 cm^{-1} but also exhibit their characteristic absorption at 1300 to 1000 cm^{-1} due to the coupling of C-O and C-C stretches. Peaks were observed at approximately 2928-

2856 and 3011 cm^{-1}, which correspond to $-CH$, $-CH_2$, $-CH_3$ and $C = C-H$ groups of the aromatic ring. These results indicated the presence of PEs on the surface of the silica gel.

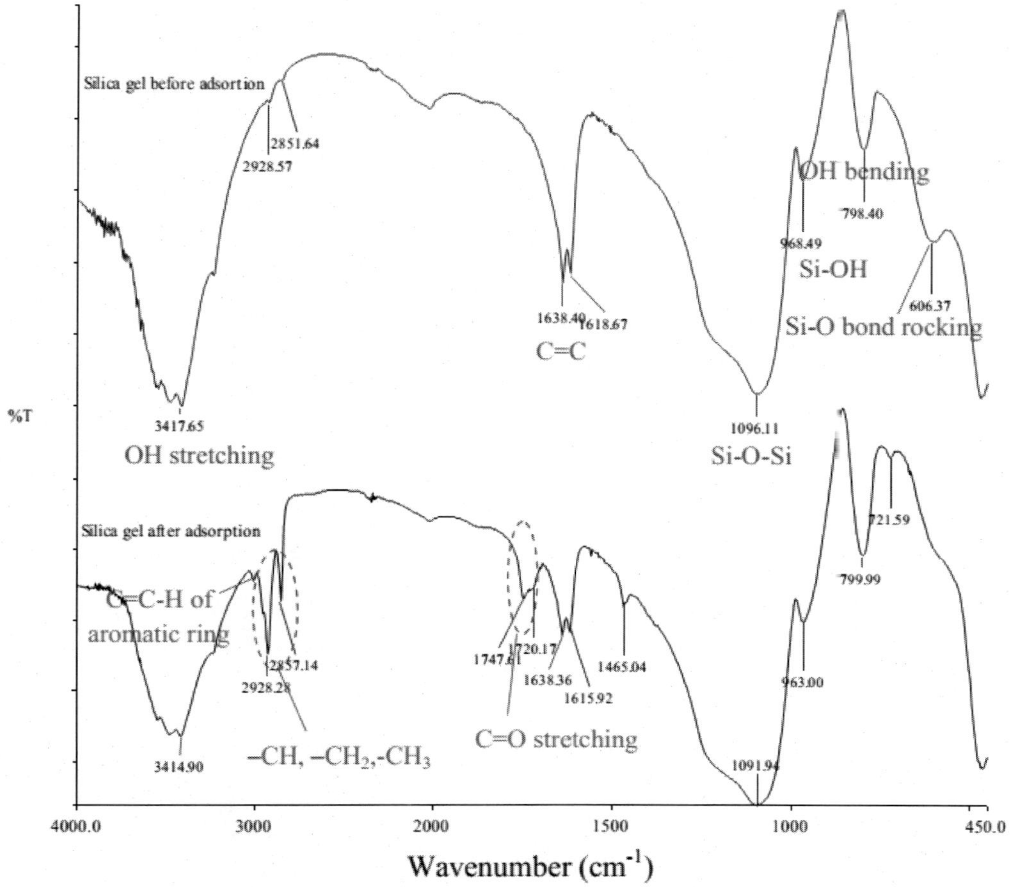

Figure 8. FT-IR Spectra of silica gel and it obtained after adsorption.

3.5. Reusability of the Silica Gel

After the adsorption process was completed, the oil was separated by filtration. The remaining silica gel in the flask was regenerated and used as the adsorbent in the next batch of adsorption tests.

The results are shown in Figure 9. A significant decreased in the percentage of PE removal and in the reusability was observed. PE removal decreased from 76.33 to 26.54%, and reusability decreased from 98.45-34.24% when the silica gel was reused for five times. Some organic compounds may remain on the surface of the recovered silica gel, as shown in Figure 10.

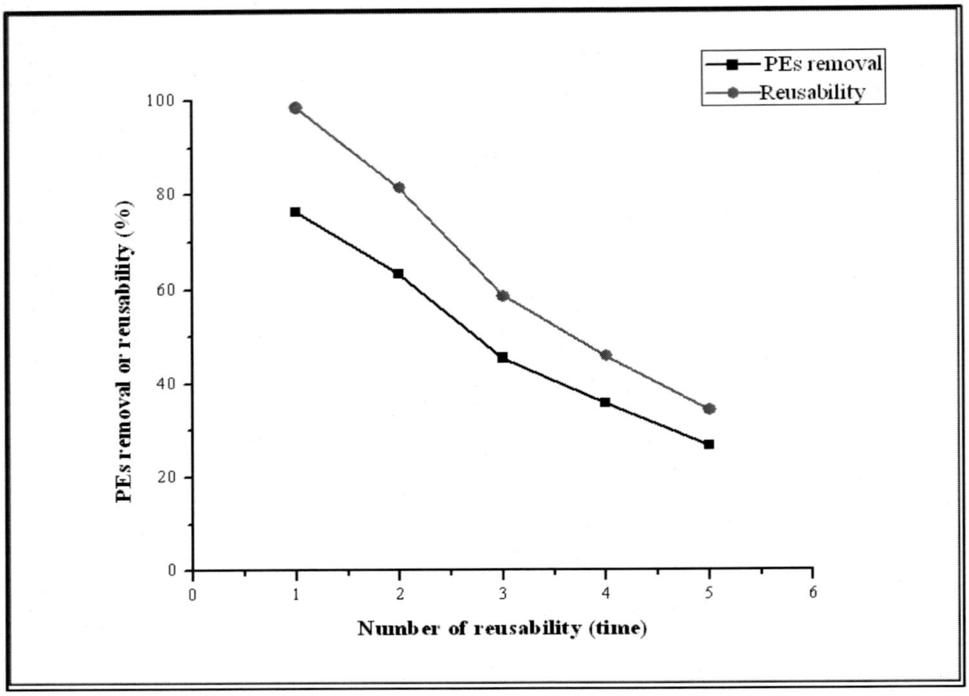

Figure 9. Percentage of PEs removal and reusability of silica gel with different time of reuse.

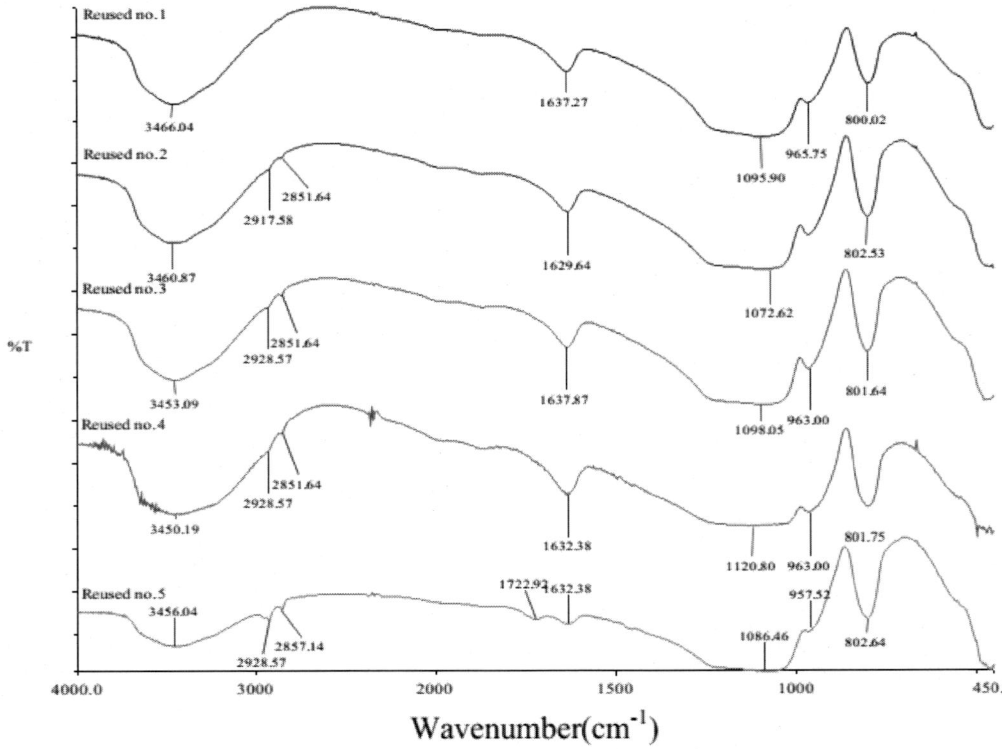

Figure 10. FT-IR spectra of the reused silica gel.

The FT-IR spectra of the first reused silica gel showed only the main functional groups of silica: the region of 3800-3200 cm^{-1}, which corresponds to the OH stretching mode of silanols (Si-OH); the peak at approximately 1094-1088 cm^{-1}, which is assigned to asymmetric Si-O-Si stretching; and peaks in the range of 970-967 cm^{-1} and 803-801 cm^{-1}, which are attributed to Si-OH bond stretching and OH bending (silanol), respectively. However, after silica gel was reused up to five times, the intensities of the C-H stretching bands in the range of 2928-2856 cm^{-1} increased. The FT-IR spectra of the silica gel reused for the fifth time shows the presence of a peak at 1722 cm^{-1}, which is attributed to the C = O of the ester; this result suggests the presence of some organic compounds on the surface of the silica gel.

The absorption intensity of the Si-OH stretching at 3456 cm^{-1} in the FT-IR spectrum of silica gel recycled five times decreased. The basic sites for recovering the adsorbent during the adsorption process after five cycles of reaction were lost. The results show that silica gel can be used two additional times for subsequent PE adsorption without significant deterioration of the percentage of reusability. A much more effective procedure for the regeneration of silica gel therefore needs to be developed to allow recycling of the adsorbent.

CONCLUSION

In this study, physical treatment using adsorption by silica gel was used to remove PE from *J. curcas* seed oil. The average PE content of the oil used was 1.68 mg TPA/g oil. The optimized conditions for PE adsorption by silica gel were studied. The experimental results show that the highest degree of PE removal (77.53%) from the oil and PE adsorption (104.18%) on silica gel were obtained when 0.3 g of silica gel/5 g of the oil was used at ambient temperature (30°C) for 60 min of adsorption and at an agitation rate of 250 rpm. A comparison of the FT-IR spectra of the silica gel and it obtained after adsorption indicates that functional groups of ester compounds were present on the surface of the silica gel obtained after adsorption. Moreover, the physico-chemical properties of the oil obtained after adsorption indicated that a brighter color of the oil without changing the fatty acid methyl ester composition was obtained. FT-IR spectra of the oil suggest that the oil was not contaminated with other compounds. The oil could be used for other applications with safer products after its adsorption onto silica gel. In addition, the reusability of the adsorbent was investigated for five additional adsorption processes. The silica gel was used two times in subsequent PEs adsorption experiments without a significant decrease in the percentage of reusability. However, a much more effective procedure for the regeneration of the silica gel needs to be developed to allow recycling of the adsorbent and to reduce the cost of PE detoxification. In conclusion, adsorption by silica gel removed PEs from *J curcas* seed oil. This method shows simplicity of design and operation and the capability to be scaled up; it also avoids the use of toxic solvents and minimizes the degradation of PEs.

ACKNOWLEDGMENTS

The authors thank Khon Kaen University for all supporting facilities and gratefully acknowledge financial support from the National Research Council of Thailand and

Rajamangala University of Technology Issan, Khon Kaen Campus, Thailand. We also especially thank to the Office of Agricultural Research and Development, Khon Kaen Province and Nakon Ratchasima Field Crops Research Center, Nakon Ratchasima Province, Thailand, for their materials support by providing *J. curcas* oil. The Center of Excellence for Innovation in Chemistry (PERCH-CIC), Commission on Higher Education, Ministry of Education, Thailand, is gratefully acknowledged for providing some chemicals used in this investigation.

REFERENCES

Anwar, J., Shafique, U., Salman, M., Zaman, W-uz. and Memoona., M. 2009.
Adsorption study of Cadmium (II) and Lead (II) on radish peels. *Journal of Scientific Research*. 1: 29-34.

Ahmed, W. A. and Salimon, J. 2009. Phorbol ester as toxic constituents of tropical *Jatropha curcas* seed oil. *European Journal of Scientific Research*. 31: 429- 436.

Aregheore, E. M., Becker, K. and Makkar, H. P. S. 2003. Detoxification of a toxic variety of *Jatropha curcas* using heat and chemical treatments and preliminary nutritional evaluation with rats. *South Pacific Journal of National Science*. 21: 50-56.

Barua, A. G., Hazarika, S., Hussain, M. and Misra, A. K. 2008. Spectroscopic investigation of the cashew nut kernel (*Anacardium occidentale*). *Open Food Science Journal*. 2: 85-88.

Belewu, M.A. and Sam, R. 2010. Solid state fermentation of *Jatropha curcas* kernel cake: proximate composition and antinutritional components. *Journal of Yeast and Fungal Research*. 1(3): 44-46.

Berghuis, N. T., Wahyuningrum, D., Zulfikar, M. A. and Soerawidjaja, T. H. 2010. Electrocatalytic hydrogenation of soybean oil using hydrogen transfer agent. In the proceedings of the Third International Conference on Mathematics and Natural Sciences, pp: 848-857.

Chumkaew, P., Karalai, C., Ponglimanont, C., Chantrapromma, K. 2003. Antimycobacterial activity of phorbol esters from the fruits of *Sapium Indicum*. *Journal of Natural Products*. 66: 540–543.

Crini, G. and Pierre-Marie, B. 2008. Application of chitosan, a natural aminopolysaccharide, for dye removal from aqueous solutions by adsorption processes using batch studies: a review of recent literature. *Progress in Polymer Science*. 33: 399-447.

Devappa, R. K., Maes, J., Makkar, H. P. S., Greyt, W. D. and Becker, K. 2010. Quality of biodiesel prepared from phorbol ester extracted *Jatropha curcas* oil. *Journal of the American Oil Chemists' Society*. 87: 697-704.

Diwani, G. E., Rafie, S. E. and Hawash, S. 2009. Antioxidant activity of extracts obtained from residues of nodes leaves stem and root of Egyptian *Jatropha curcas*. African *Journal of Pharmacy and Pharmacology*. 3: 521-530.

Donlaporn, S. and Worapot, S. 2011. Toxic compound, Anti-nutritional factors and functional properties of protein isolated from detoxified *Jatropha curcas* seed cake. *Molecular Science*. 12: 66-77.

Goel, G., Makker, H.P.S., Francis, G. and Becker, K. 2007. Phorbol Esters: Structure, Occurrence and Biological Activity. *International Journal of Toxicology*. 26: 279–288.

Haas, W., and Mittelbach, M. 2000. Detoxification experiments with the seed oil from *Jatropha curcas* L. *Industrial Crop and Products*. 12: 111–118.

Joshi, C., Mathur, P. and Khare, S. K. 2011. Degradation of phorbol esters by *Pseudomonas aeruginosa* PseA during solid-state fermentation of deoiled *Jatropha curcas* seed cake. *Bioresource Technology*. 102: 4815-4819.

Makkar H. P. S., Becker K., Sporer F., Wink M. 1997. Studies on nutritive potential and toxic constituents of different provenances of *Jatropha curcas*. *Journal of Agricultural and Food Chemistry*. 45: 3152-3157.

Makkar, H. P. S., Aderibigbe, A. O. and Becker, K. 1998. Comparative evaluation of non-toxic and toxic varieties of *Jatropha Curcas* for chemical composition, digestibility, protein degradability and toxic factor. *Food Chemistry*. 62: 207-215.

Makkar, H. P. S. and Becker, K. 2009. *Jatropha Curcas*, A promising crop for the generation of biodiesel and value-added coproducts. European Journal *of* Lipid Science *and* Technology. 111: 773–787.

Martínez-Herrera, J., Siddhuraju, P., Francis, G., Dávila-Ortíz, G. and Becker, K. 2006. Chemical composition, toxic/antimetabolic constituents, and effects of different treatments on their levels, in four provenances of *Jatropha curcas* L. from Mexico. *Food Chemistry*. 96: 80-89.

Parida, S. K., Dash, S., Patel, S. and Mishra, B. K. 2006. Adsorption of organic molecules on silica surface. *The Journal of Colloid and* Interface Science. 121: 77-110

Rakshit, K. D., Darukeshwara, J., Raj, K. R., Narasimhamurthy, K., Saibaba, P. and Bhagya, S. 2008. Toxicity studies of detoxified *Jatropha* meal (*Jatropha curcas*) in rats. *Food and Chemical Toxicology*. 46: 3621-3625.

Sasithorn, J., Wiwattanadate, D. and Sangsuk, S. 2010. Utilization of fly ash from power plant for adsorption of hydrocarbon contamination in water. *Journal of Metals, Materials and Minerals*. 20: 5-10.

Siddhuraju, P., Makkar, H.P.S., Becker, K. 2002. The Effect of ionising radiation on antinutritional factors and the nutritional value of plant materials with reference to human and animal food. *Food Chemistry*. 78: 187–205.

Tuberoso, C. I. G., Kowalczyk, A., Sarritzu, E., Cabras, P. 2007. Determination of antioxidant compounds and antioxidant activity in commercial oil seeds for food use. *Food Chemistry*. 103: 1494-1501.

In: Jatropha Curcas
Editor: Gregory Medina

ISBN: 978-1-63483-089-8

Chapter 9

RESPONSES OF TWO *JATROPHA CURCAS* L. PROVENANCES ACCORDING TO WATER AND NITROGEN AVAILABILITIES

Sameh Cherif, Zouheir Nasr and Mohamed Larbi Khouja
Laboratory of Ecology and Improvement Sylvo-Pastoral
National Institute for Research in Rural Engineering,
Water and Forestry, INRGREF, Ariana, Tunisia
Faculty of Sciences of Bizerte, University of Carthage,Tunisia

ABSTRACT

This study aims to control the combined effect of water stress and nitrogen deficiency on growth and physiology of plants of *Jatropha curcas*, aged one year and issued from two different origins provenances: PAR (Paraná-Brazil), SUR (Surinam-North South America). The water deficit was applied during the month of October 2011. Two water regimes were adopted (I: irrigated plants, NI: non irrigated plants) associated with two nitrogen regimes (N-: no nitrogen, N+:2 g of nitrogen). Some physiological parameters were measured (leaf chlorophyll content, soil moisture and relative water content on leaf) as well as growth parameters (Stem diameter of the stem and leaf area, leaf thickness) were monitored during this survey. The results revealed that the lack of water and nitrogen decrease the averages of stem diameter and the leaf area: This was more emphasized on Surinam provenance. We noticed also an increase on leaf thickness especially on Paraná provenance. Furthermore, the ratio root biomass / aboveground biomass had increased more with Paraná provenance; this ratio indicated best tolerance to drought. Chlorophyll content in the leaf and water content in the soil are very sensitive to the combined deficit of water and nitrogen: These were significantly reduced on Surinam provenance. However, the relative water content of leaves were not often affected by water deficit especially on Paraná provenance, and were generally maintained at relatively high values (>50%).

Keywords: Jatropha, water, stress, nitrogen, physiology, morphology

INTRODUCTION

The growing interest in *Jatropha curcas* species is explained not only by its ability to grow on marginal lands [13] but also by its abilityZouheir Nasr to fight erosion and its important role in rural development for some African countries [1]. Tunisia is among countries that are interested in Jatropha for its energy, agronomic and ecological benefits. Successfully introduced in tropical countries, particularly India, the adaptation of *Jatropha curcas* to the Mediterranean climate and more specifically in Tunisia, is still unknown.

Jatropha is a shrub that grows in tropical and subtropical regions in Latin America [14, 16]. Its seeds contain 27% to 40% of oil, which has been long used in traditional medicine for livestock feed or making soap. Recently, Jatropha is used to produce agro fuels [1]. Hence, it is nicknamed "green gold of the desert."

For this study, we aim to compare the response of lack of water and nitrogen on the development of two provenances of *Jatropha curcas* L. These provenances are planted in the nursery of the National Institute for Research in Rural Engineering Water and Forestry. The objectives of this study are three (i) compare the response of two provenances according to water stress and nitrogen deficiency, (ii) identify the agro-climatic zone for each provenance, (iii) assess the need for irrigation and fertilization cultivation of *Jatropha curcas* in Tunisia.

MATERIELS AND METHODS

A. Overview of the Test

Two provenances of *Jatropha curcas* L. had been the subject of our study: one was from Brazil (Paraná) with the code PAR and the other was from northern South America (Surinam region) which code is SUR. The seeds were provided by the Italian Society "Agroils" through a Tunisian-Italian bilateral project. They were planted on July 2011 in plastic bags (black polyethylene) 2-liter volume at the nursery of the INRGREF in a compost-based substrate (sand mix 1/3 soil and 2/3). The seeds were buried (one per cell) at a depth of 1cm. After planting the seeds, the irrigation was performed.

B. Nitrogen-water Treatments Administration

The plants were arranged in four blocks of 16 pots each, following an organizational split-plot treatment combining water and nitrogen treatment, to obtain the following combinations:

I+/N+: Irrigated with nitrogen
I+/N-: Irrigated without nitrogen supply
I-/N+: Not irrigated with nitrogen supply
I-/N-: Not irrigated without addition of nitrogen

The irrigated plants were watered manually every day at 9 o'clock.

C. Measurement of Morphological Parameters

The various morphological parameters: leaf thickness, collar diameter ar d leaf area were measured at the beginning and the end of each drying cycle.

1. Leaf thickness

This parameter was measured through the volume ratio of leaves in watei / leaf area.

2. Collar diameter

Collar diameter was measured at the beginning and the end of each drying cycle using a tape meter.

3. Leaf area

Leaf area was measured using a planimeter Windias type Delta-T Devices LTD, which allows advanced analysis and is fast compared to other types. The different calculated parameters are leaf area, perimeter, width and length.

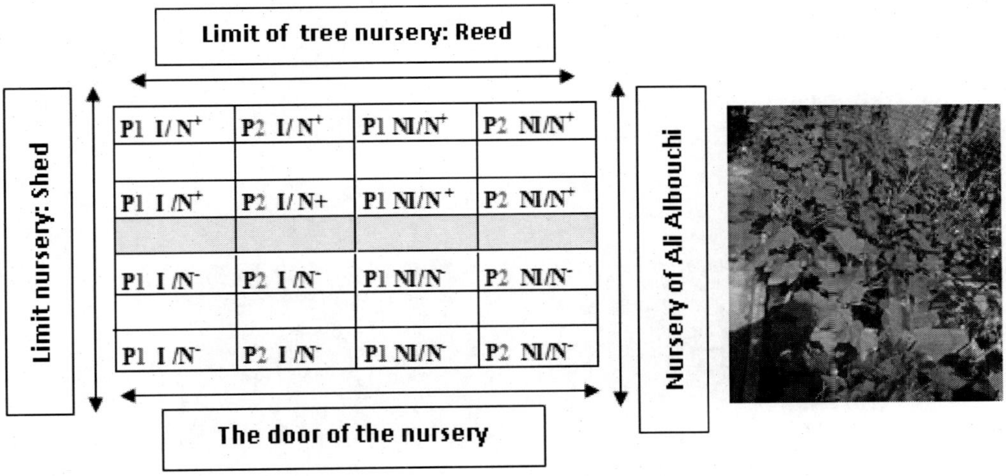

Figure 1. Experimental set up at the nursery of the INRGREF, October 2011 (P1: Provenance PAR, P2: Provenance SUR, I+: Irrigated, I-: No irrigated, N+: With nitrogen, N-: Without nitrogen).

D. Measurement of Physiological Parameters

1. Relative Water Content

The relative water content of the leaf was determined by the method described by (Schonfeld and al, 20). Under this method, the leaves are cut at the base of the blade; they are weighed immediately for fresh weight (Fw). These leaves are then put into test tubes filled with distilled water and placed in the dark in a cool place. After 24 h, the leaves are removed, placed in a paper to absorb water from the surface, weighed again to obtain the turgid weight (Tw).The samples are then placed in an oven set at 70°C for 72h and weighed for their dry weight (DW).

Figure 2. Planimeter windias.

The relative water content is calculated by the following formula:

RWC (%) = $(Fw\_D_W) / (Tw—D_W)$

RWC: relative water content
Fw: fresh weight
D w: dry weight
Tw: turgid weight

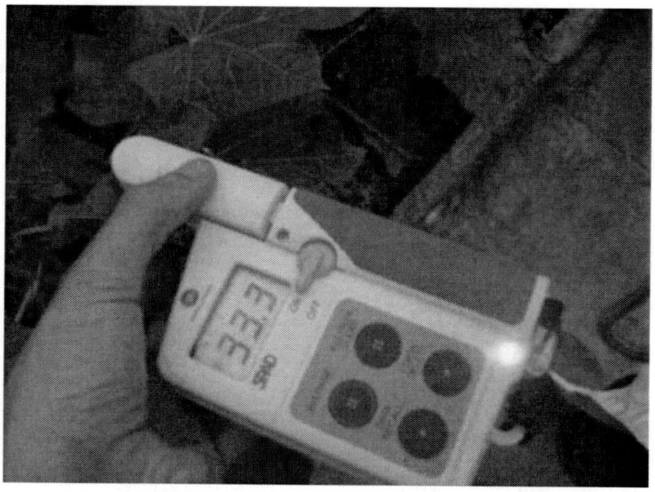

Figure 3. Chlorophyll meter SPAD-502.

2. Content of Chlorophyll

The SPAD chlorophyll meter is a portable tool diagnostic that measures the relative content of leaf chlorophyll. The SPAD (Soil Plant Analysis Development) uses the reflectance ratio between red (650nm) and infrared (950nm). Repeated use of chlorophyll meter throughout the cycle may indicate at an early time the approach of a nitrogen deficiency. This helps to avoid any loss of performance [4].

3. Volumetric Water Content

The measurement of volumetric water content in soil was determined by using TDR 300 (Time Domain Reflectometry). This device measures the dielectric permittivity to estimate soil moisture in all conditions (dry to saturation).

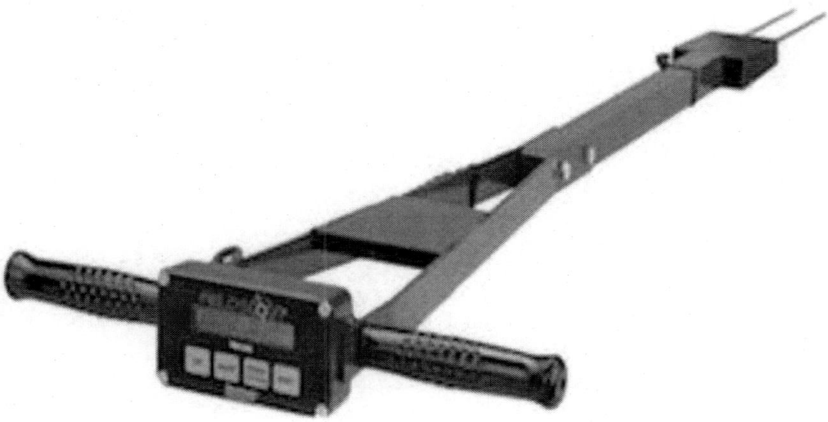

Figure 4. The Soil Moisture Field Scout TDR 300.

E. Statistical Analyses

The results were subject to analysis of the variance of two factors through the GLM (General Linear Models) of SAS. The averages are compared using the Newman-Keuls test at 5% risk threshold.

RESULTS AND DISCUSSION

A. Effect of Water and Nitrogen Deficit on Morphological Parameters of Plants

1. Leaf Area

Leaf area is dependent on irrigation and nitrogen fertilization, in fact it reaches a maximum value for the two provenances: 112.81cm^2 from PAR; 101.62 cm^2 from SUR) when plants are irrigated with a nitrogen supply I/N+.

Figure 5 shows that the average of leaf area from provenance PAR 112.81 cm^2 are higher than the leaf area average observed on the origin SUR 101.62 cm^2.

On the one hand, leaf area decreased more for the provenance SUR with severe water deficit and lack of nitrogen: The difference between the treatment NI/N- and I/N- was 14.51cm2.

On the other hand, this difference is only 8.64 cm^2 for the provenance PAR. The effect of the irrigation seems to be much more important for the SUR provenance. Moreover, the nitrogen had no effect on leaf area in case of water deficiency. Indeed, the difference between plants NI/N- and NI/N+ was 3.38 cm^2 for the provenance SUR, against 1.69cm^2 for the provenance PAR.

These statistical analysis results suggest a significant effect of water and nitrogen uptake on leaf area ($p < 0.05$), whereas there were no difference between provenances on this parameter ($p = 0.1097$).

2. Leaf Thickness

The combined deficit in water and nitrogen caused an increase in leaf thickness which increased by 0.47cm for SUR and 0.48cm for PAR (Figure 6). The statistical analysis showed a significant effect of both treatments on the leaf thickness while the effect of provenance was not significant ($p = 0.2618$).

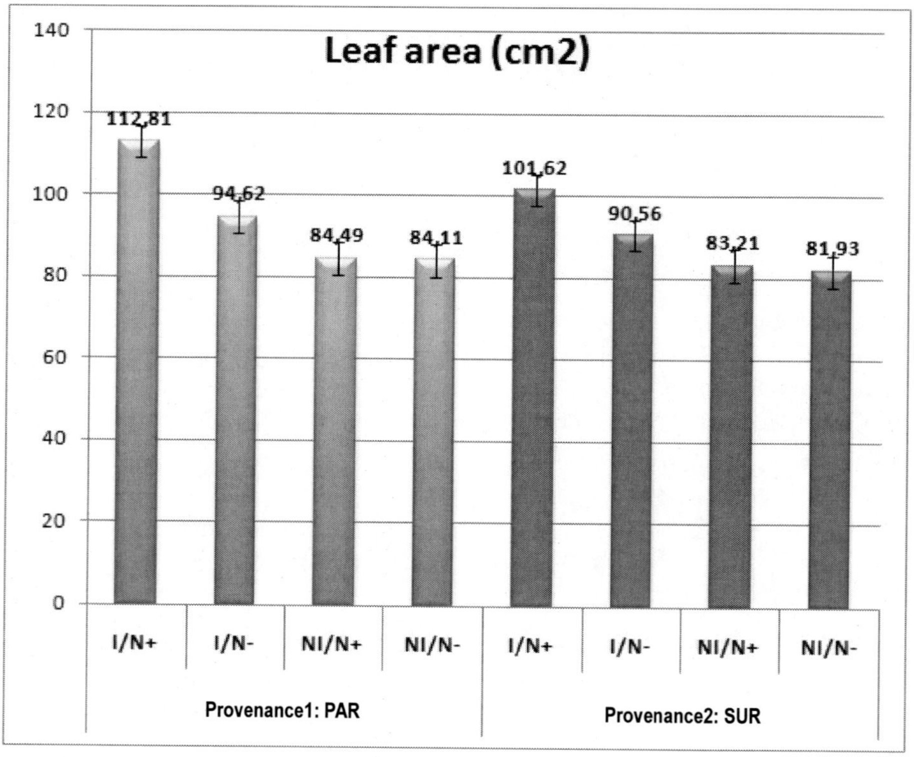

Figure 5. Effect of different treatments on leaf area of two different provenances of Jatropha (Adjusted Mean ± standard error).

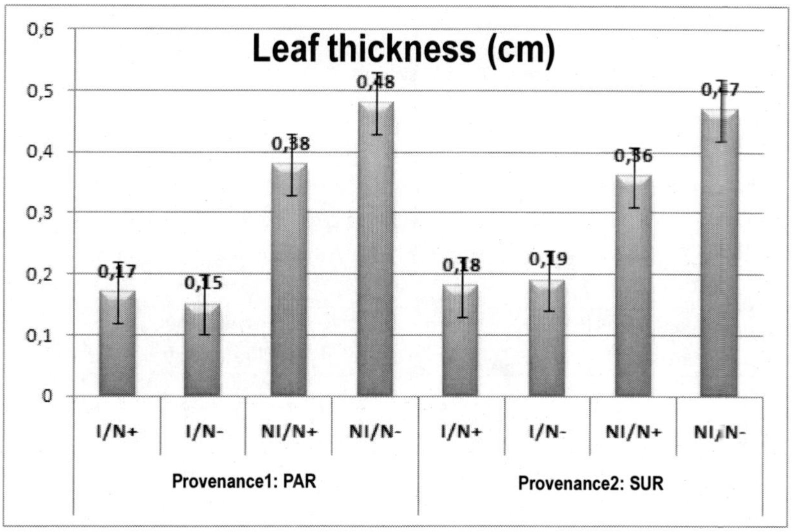

Figure 6. Effect of various treatments on the leaf thickness I/N+, I/N-, NI/N+, NI/N-(Adjusted Mean ± standard error).

3. Relationship between Root Biomass and Aboveground Biomass

The ratio between root biomass and aboveground biomass increased significantly for both provenances when plants are submitted to water stress and nitrogen deficiency (NI/N-). We noted a slight increase for the PAR provenance (Figure 7).

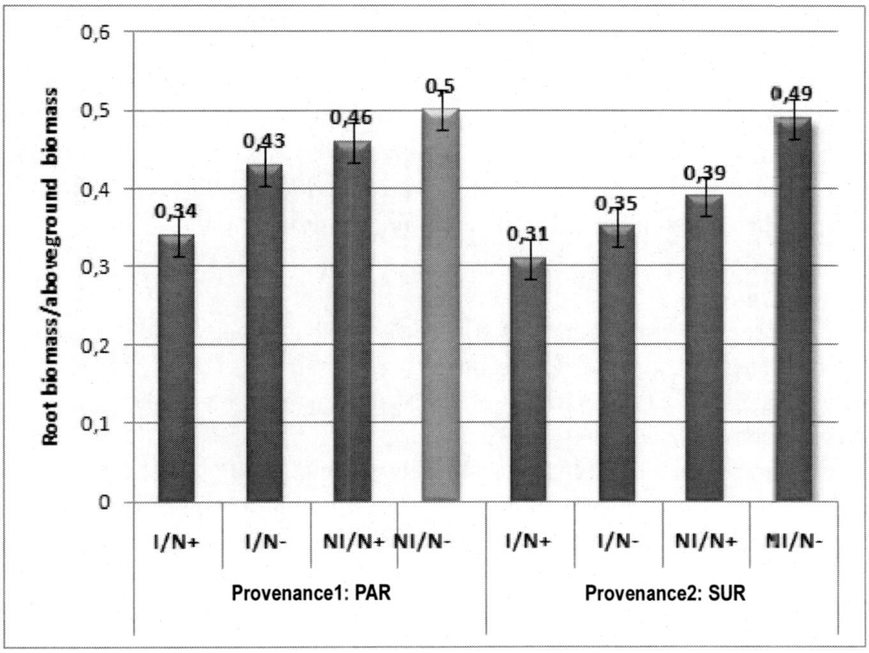

Figure 7. Consequences of water and nitrogen deficiencies on relative root biomass/aboveground biomass in two provenances of *Jatropha curcas L.* Adjusted mean ± standard error.

4. Stem Basal Diameter

The stem basal diameter decreased without irrigation and nitrogen fertilization; it increased by 6.13 cm to 3.20 cm for the SUR provenance while with the PAR provenance, it varied between5.75and 3.30cm. (table 1).

Table 1. Effect of different treatments on diameter at the base of the stem in two provenances of *Jatropha curcas* L
*** The unit of diameter at the base of the stem is cm**

	Provenance :PAR Treatment				Provenance :SUR Treatment			
Date	I/N^+	I/N^-	NI/N^+	NI/N^-	I/N^+	I/N^-	NI/N^+	NI/N^-
03/10/11	4,54	4,4	4,5	4,44	4,55	4,47	4,48	4,35
13/10/11	4,9	4,56	4,45	3,66	5,46	5,01	4,43	4,01
24/10/11	5,75	5,30	4,08	3,3	6,13	5,71	3,9	3,2

B. Effects of Water and Nitrogen Deficit on Physiological Parameters

1. The Soil Water Content

During the first and the second drying cycles, the volumetric soil water content has declined for non-irrigated plants and without nitrogen supply (NI/N^-) in both provenances (Figure 8). This is very important in the second drying cycle on the SUR provenance: the difference between (NI/N^-) and (I/N^-) was 23%, whereas on the PAR provenance it was only19%.

With the addition of nitrogen and under a water-free (NI/N+), water reserves in the soil were not significantly different from those submitted to treatments (NI/N-). This result comply with the statistical analysis as it confirmed that the supply of nitrogen has no influence on water content in the soil for the two provenances $p = 0.8476$ but the provenance and water supply had a significant effect on the soil moisture ($p < 0.05$).

2. Chlorophyll Content

The leaf chlorophyll content increased especially when plants were irrigated and with nitrogen supply. The lowest value was recorded for plants NI/N- (table 2). Furthermore, the analysis of variance showed a significant effect of provenance, the availability of water and nitrogen on chlorophyll content ($p < 0.05$).

A slight difference in chlorophyll content between plants NI/N- and NI/N+ emphasizes the insignificant effect of nitrogen with dryness. The chlorophyll content increased by14.81% to 16.97% for SUR provenance and by 13.75% to 15.11% for PAR provenance at the end of the drying cycle. The provenance SUR needs more nitrogen to its photosynthetic activity than the PAR provenance. The deficit of water influenced significantly the chlorophyll content highlighted by the data in Figure 9.

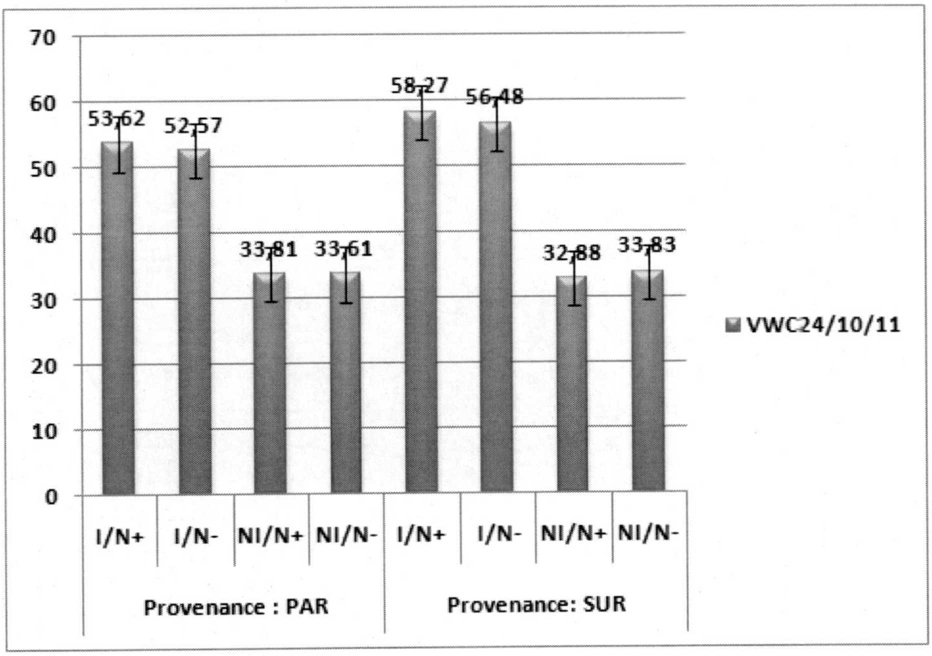

Figure 8. Variation of volumetric water content in soil (VWC) on the date 24/10/2011.

Table 2. Evolution of chlorophyll content measured by SPAD for different treatments (I/N+: irrigated plants with nitrogen supply, I/N-: irrigated plants without nitrogen supply, NI/N+: non irrigated plants with nitrogen supply, NI/N-: non-irrigated plants without nitrogen)

	Provenance :PAR Treatment				Provenance :SUR Treatment			
Content of chlorophyll(%)	I/N+	I/N-	NI/N+	NI/N-	I/N+	I/N-	NI/N+	NI/N-
SPAD03/10/11	32,26	31,45	30,1	28,04	34,1	34,09	33,34	31,45
SPAD 13/10/11	33,73	30,67	27,75	24,11	35,62	34,53	30,42	26,12
SPAD 24/10/11	33,83	33,28	15,11	13,75	36,51	34,98	16,97	14,81

3. Correlations between Water Availability and Chlorophyll Content

The major role of water and nitrogen, in improving the photosynthetic activity, was emphasized by data in Figure 9. It revealed a strong correlation between soil water reserves and the chlorophyll content of leaves for all treatments. This correlation appeared to be dependent on the intensity of water stress in plants. Indeed, the chlorophyll content increased more with the plants that were irrigated and with nitrogen supply for the two provenances. However the chlorophyll content decreased when plants were deprived of water and nitrogen.

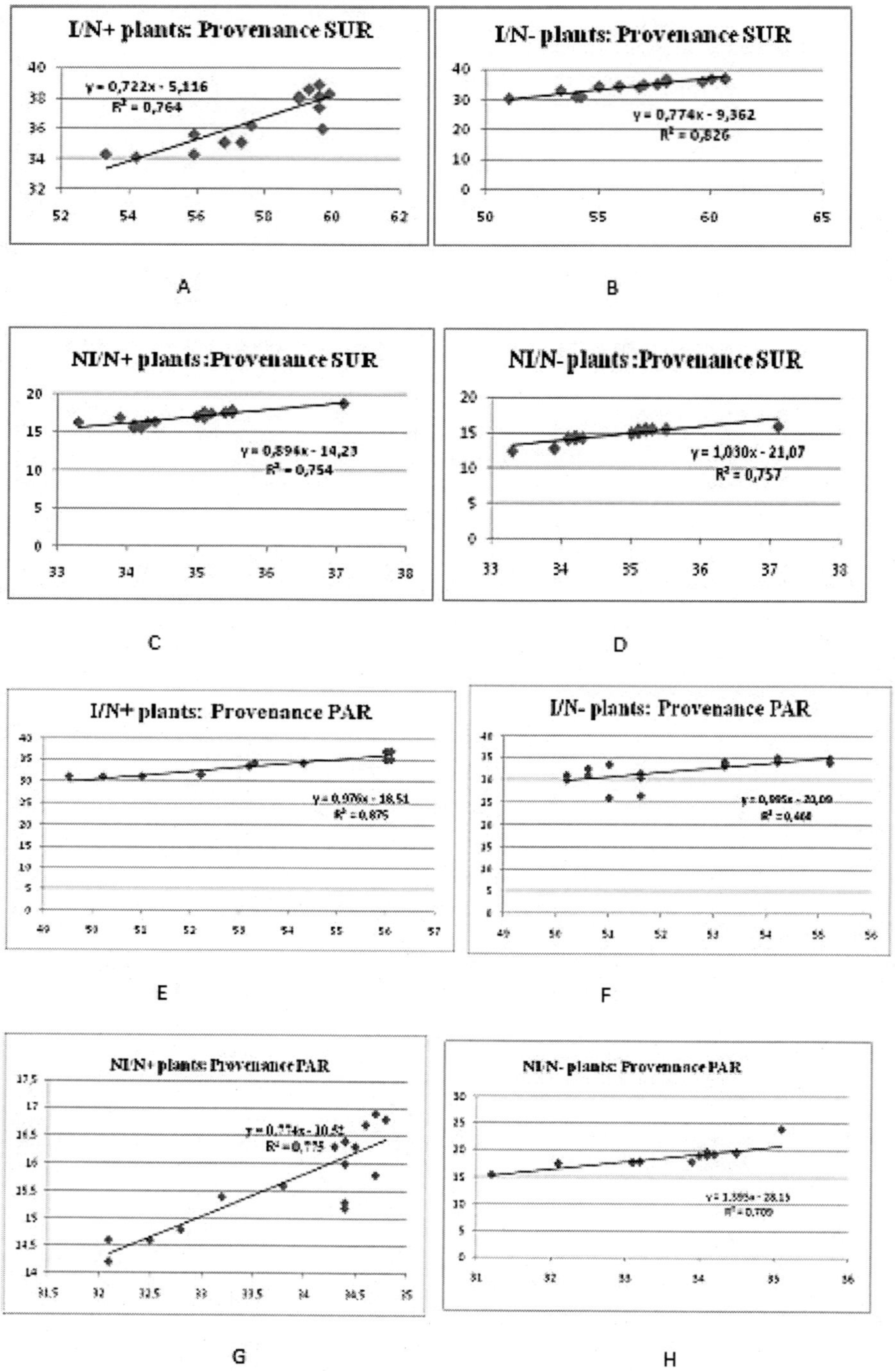

Figure 9. Different correlation between soil water content and chlorophyll content for all types of treatments applied to the two provenances of the plant *Jatropha curcas* L. at the end of drying cycle.

(A, B, C and D): SUR provenance.

(E, F, G and H): PAR provenance.

4. Relative Water Content of Leaves

During the drying period, leaf water content decreased due to the lack of water and nitrogen (NI/N-) for both sources, but this decrease is larger for SUR provenance. By comparing this treatment with I /N+, a difference of 16.05% was noted; whereas it was 11.48% for the PAR provenance.

This is presented in Figure 10. Statistical analysis showed a significant effect of water uptake on the turgidity of leaf, however the addition of nitrogen did not influence the water content of the leaf level p = 0.0789.

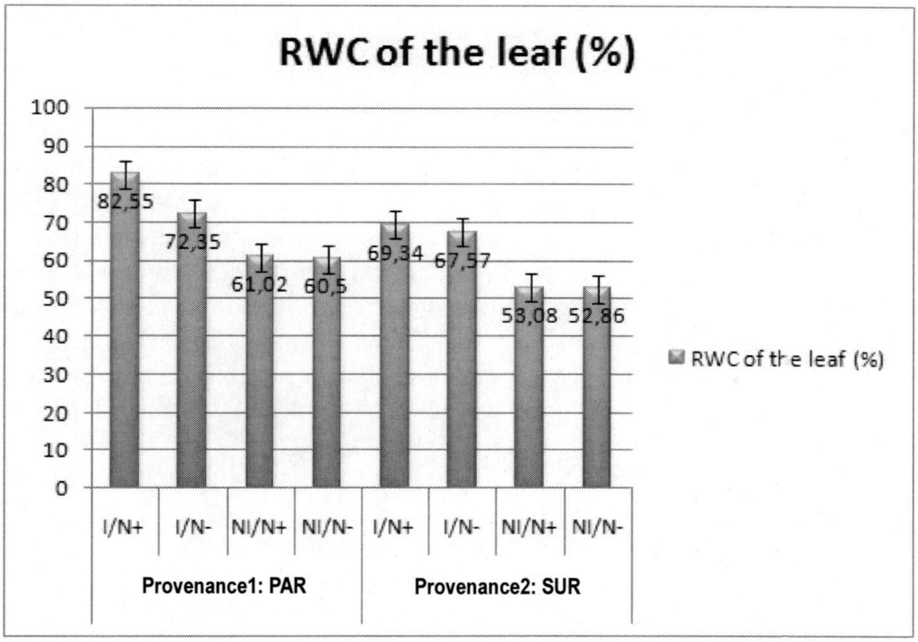

Figure 10. Response of relative water content of the leaves level with water and nitrogen deficit (Adjusted Mean ± standard error).

DISCUSSION

A. Morphological Behavior of the Two Provenances

Our study showed that in case of deficiency of water and nitrogen, leaf area decreased significantly by 32.7 cm^2 for SUR provenance while it decreased slightly by 19.7cm^2 for PAR provenance.

Indeed, with a deficiency of water and nitrogen each provenance adjusted its response by a reduction of leaf area, leading to a reduction in the amount of water through transpiration [5]. This strategy was mentioned by several authors who showed that the reduction in leaf area is an adaptation mechanism to drought [17, 19]. Moreover it helps to maintain hydration of the tissues more or less suitable to the extreme conditions of water stress and nitrogen. However, a decrease in leaf area can affect the performance of plants because of the reduction of photosynthetic capacity. Hopkins, [12] and Manivannan [18] found that water deficit causes a delay in plant growth that can be considered as a form of plants adaptation to reduce the size of their aerial biomass and develop their underground one [9].

The increase in leaf senescence reduced aboveground biomass and may subsequently affect physiological mechanisms mainly the reduction of transpiration and the rate of CO2 fixation by photosynthesis via a stomatal closure [6]. We noted a decrease in diameter at the base of the stem level of plants NI/N-. This decrease was more important for the SUR. Nevertheless, a slight decrease compared to the control plants (I/N+) was noticed for the PAR provenance. These results comply with the results found by Achten, [2] which showed that the stem diameter decreased after 12 days of severe water stress.

The combined effect of lack of water and nitrogen could lead to cutinisation of the leaves. Indeed, the leaves are thicker. This change in the leaves may explain the reduction in net photosynthetic rates of plants [17].

Water deficit and nitrogen reduce aboveground biomass; the highest values are affected by a decrease recorded in the SUR provenance although the PAR provenance seems to be less affected and more tolerant to this deficit. The same results found by Gonzalez [11] showed that the total biomass of the plant depends on the intercepted photosynthetic radiation.

The nitrogen has an important role in the growth and the development of biomass, it helps to improve the vegetative growth of seedlings of *Jatropha curcas* and increase their biomass and leaf area which is consistent with the results of Ksontini [15] on pedunculate oak. According to Albouchi [3] the growth of root is considered as criteria for drought tolerance. In our case the PAR provenance is most tolerant to drought and nitrogen deficiency.

B. Physiological Behavior of the Two Provenances

The water content in soil was reduced in case of lack of water and nitrogen. This reduction affects more the SUR provenance. Indeed, the difference between NI/N -and I / N + was 24, 44% while this difference is only 20, 01% for the PAR provenance. The decrease in water availability led to a drying of the soil and in turn caused a change in plant water status, indicated by the decrease in relative water content of leaves. Several experiments pointed out that the response of the plant through stomatal closure is more related to soil moisture than the leaf water status. The consequence is a decrease in chlorophyll content and photosynthetic activity [8].

In addition, the results obtained demonstrate the close relationship between chlorophyll content in the leaves and the content of the soil with water and nitrogen. We note that when the ground receives a supply of water and nitrogen, chlorophyll content is high. The degree of water stress also affects chlorophyll content and in the case of severe water stress and without nitrogen supply, chlorophyll content decreases more for SUR provenance compared to the PAR provenance.

The decrease of chlorophyll content in case of water deficit and nitrogen couldn't be mainly due to stomatal factor but also to photosynthesis systemII. Despite the decline in RWC the plant keeps high levels at the end of drying cycle.

The water content in the leaf level was maintained at high values relatively (> 50%). The osmotic adjustment allows the accumulation of osmotic solutes in the vacuole resulting in a decrease of cellular osmotic potential, which leads to maintaining cell turgor [10, 17].

The increase of biomass roots is considered as an indicator of survival to dehydration which is more noticed in the PAR provenance. Our results show also a close relationship between the water content in soil and chlorophyll content in the leaf. The PAR provenance is less sensitive to the need of water supply and nitrogen fertilization than the SUR provenance.

Furthermore, the relative water content in leaves decreases when plants are submitted to water stress and lack of nitrogen but it keeps a fairly high value for PAR provenance compared to the SUR provenances which indicate a possible osmotic adjustment.

CONCLUSION

Despite the different origin of two *Jatropha curcas* provenances, they behave similarly towards a lack of water and nitrogen with a slight resistance for the PAR provenance. In fact, the two provenances can not handle very long dry periods and nitrogen deficiency. According to results, *Jatropha curcas* L. can be grown in humid or sub-humid bioclimatic zones with additional irrigation in summer or in semi-arid climate with full irrigation (between April and October) with a preference for the PAR provenance. In filter soil and without nitrogen, a fertilizer is required for the growth. It seems difficult to cultivate Jatropha in arid or desert floor in Tunisia.

REFERENCES

Abdrabbo A. Abou Kheira, Nahed M.M. Atta 2009.Response of *Jatropha curcas* L. To water deficits: Yield, water use efficiency and oilseed characteristics. *Biomass and Bioenergy.* 33: 1343-1350.

Achten, W.H. Maes,B. Reubens, E. Mathijs, V.P. Singh, L. Verchot, B. Muys 2010. Biomass production and allocation in *Jatropha curcas* seedlings under different levels of drought stress. *Biomass and Bioenergy.*

Albouchi A 1997.Effets du pré conditionnement à la carence eneau sur certaines manifestations de l'endurcissement à la sécheresse chez Acacia cyanophylla Lind I. Croissance, comportement hydrique et accumulation active de solutés. *Thèse de Doctorat, Fac.Sci.Tunis,* 142p.

Alena Torres Netto, Eliemar Campostrini Jurandi Gonçalves de Oliveira, Ricardo Enrique Bressan-Smith2004.Photosynthetic pigments, nitrogen, chlorophyll a florescence and SPAD-502 readings i cofee leaves Sientia *Horticulturae* (2005).104: 199-209.

Alves A.C.C et Setter T.L.2000. Response of cassava to water deficit: leaf area growth and abscisic acid. *Crop Sci,* 40: 131-137.

Chaves M.M., Maroco J.P.et Pereira J.S. (2003).Understanding plant responses to drought from genes to the whole plant. *Funct. Plant Biol,* 30: 239-264.

Davies W.J., Tardieu F. et Trejo C.L. 1994. How do chemical signals work in plants that grow in drying soil. *Plant Physiol.* 104:309- 14.

Davies WJ., Zhang J. (1991) Root signals and the regulation of growth and development of plants in drying soil. *Plant Physiology,* 42:55-76.

Dichio, Romano M., Nuzzo V., Xiloyannis C.2002.Soil water availability and relationship between canopy and roots in young olive trees (cv coratina). *Acta Hortic. 586*:255-258.

El Jaafari S., 2000. Durum wheat breeding for abiotic stresses resistance: Defining physiological traits and criteria. *Option Méditerranéenne,* 40: 251-256.

Gonzalez D. (2006).Effet du déficit hydrique sur l'état de nutrition azotée chez les graminées fourragères. *Thèse de doctorat. Université de Poitiers.* P164.

HopkinsW.G.2003.Physiologie végétale.Traduction de la2iéme édition américaine par SergeRambour.Révision scientifique de Charles. *Marie Evrardd.*452-459.

Kheira A, Atta N.2009. Response of *Jatropha curcas* L. to water deficit: yield, water use efficiency and oil seed characteristics. *Biomass Bioenergy* 33:1343.

Kochhar S, Singh S, Kochhar V. 2008. Effect of auxins and associated biochemical changes during clonal propagation of the biofuel plant *Jatropha curcas. Biomass Bioenergy*. 32.

Ksontini, M., Louguet, P., Laffrey, D. & Rejeb, MN. (1998). Comparaison des effets de la contrainte hydrique sur la croissance, la conductance stomatique et la photosynthèse de jeunes plants de chêne méditerranéens (Quercus suber, Quercus faginea, Qurcus coccifera) en Tunisie. *Annals of forest science,* 55: 477-495.

[Kumar, A., Sharma S., 2008. An evaluation of multipurpose oil seed crop for industrial uses (*Jatropha curcas* L.): A review. *Industrial Crops and Products. En cours d'impression.*

Lefi, E., Ben Younes, Medrano, H. 2004. Seasonal variation of leaf production for Medicago arborea and Medicago citrine under irrigation and drought conditions. *Options Méditerranéenne.* 38: 327-330.

ManivannanNPuppala, N. et Delikostadinov S.G.2007.Association between pod and Kernel characteristics in Valencia groundnut Arachis hypogaea L.subsp. Fastigiata Var. Fastigiata. *J. Oilseeds Res.* 24: 170-171.

SantestebanL.G., Miranda C.et Royo J.B.2009.Effect of water deficit and rewatering on leaf gas exchange and transpiration decline of excised leaves of four grapevine (Vitis vinifera L.) cultivars. *Scientia Horticulturae.* 121: 434- 439.

Schonfeld M.A., Johnson R.C., Carver B.F., et Mornhinweg D.W.1988.Water relations in winter wheat as drought resistance indicator. Cr*op Sci;* 28: 526-531.

Slatyer R.O (1974) Effects of short periods of water stress on leaf photosynthesis. In: *Plant responses to climatic factors, R.O. Slatyer (ed),* pp 271-276.Paris, Unesco.P.

In: Jatropha Curcas
Editor: Gregory Medina

ISBN: 978-1-63483-089-8

Chapter 10

MECHANICAL PROPERTIES OF WHITE BREAD MADE OF BARLEY FLOUR CONCENTRATE AND *JATROPHA CURCAS*

***N. Guemes Vera*[1,*] *and Alfonso Totosaus Sanchez*[2]**
[1]Centro de Investigación en Ciencia y Tecnología de Alimentos, Instituto de Ciencias Agropecuarias, Universidad Autónoma del Estado de Hidalgo, Tulancingo, México
[2]Food Science Lab, Tecnologico de Estudios Superiores de Ecatepec, México, Estado de México, México

ABSTRACT

Mexico is a country with a great variety of natural resources like seeds and grains that can be employed to improve bread manufacture. Recent investigations shown that *Jatropha* flour can be used to improve bread nutritional properties. The aim of this work was to obtain optimal mixtures of wheat flour, barley concentrate and *Jatropha curcas* flour, to develop a bread product (white bread), evaluating doughs and breads mechanical properties. *J. curcas* and barley concentrate doughs mechanical evaluation indicated that the optimum formulation to bread formulation was with 3.7% of both ingredients. The hardness, adhesiveness, cohesiveness and elasticity of this bread were lower than the control bread. Control bread crumb quality presented no-homogeneous and more open alveolus, as compared with the closer and more uniform crumb in the formulated bread, with lower loaf volume. Higher fiber content in barley concentrate resulted in higher bread weight, with higher water retention capacity. Proximal chemical analysis showed a protein content of 11.05 and 7.47 of fiber.

* Corresponding author: njgv2002@yahoo.com.mx.

INTRODUCTION

Foods made with wheat are incomplete protein sources. This proteins contain all 8 essential amino acids but they do not have appropriate levels, so the combination of wheat with other foods provides could be correct. However when compared with other cereals such as rice and corn had have less protein that wheat (Callejo, 2002; Desrosier, 1999; Belitz y Grosch, 2000). Gluten proteins, representing approximately 80-90% of total protein from white flour, these proteins show baking quality of flour (Callejo, 2002; Calaveras, 1996).

Gluten contains essentially two groups of proteins:

- Gliadin: Defined as the fraction insoluble in saline solutions.
- Glutenins: Insoluble in saline solutions and alcoholic solutions.

Gliadins and glutenins are synthesized only in the developing endosperm of the wheat kernel; they have similar amino acid compositions, but there are major chemical and physical differences among them, they are of particular relevance in their functional characteristics (Callejo, 2002).

The Main Functions of Gluten in Baking are:

- Water absorption during kneading.
- Transmits the doughs physical properties: extensibility and elasticity.
- It is responsible loaf volumen and the texture.

Barley, monocot is an annual plant belonging to the family Poáceas (grasses), is represented by two major crop species: *Hordeum distichon* L., which is used for obtaining beer, and *Hordeum hexastichon* L., which is basically used as fodder for animal feed; both species can be grouped under the single name *Hordeum vulgare* L. (Czuchajowska et al., 1992; Szczodrak et al., 1992).

It is an excellent source of carbohydrates, which constitutes about 80% of grain weight (Czuchajowska et al., 1992; Szczodrak et al., 1992). It is contains high levels of β-glucan, This is contributors of dietary fiber, a crucial component of the human diet (Newman y Newman 1991, Granfeldt et al., 1994). Starch is the most important component in barley, representing up to 65% of the dry weight of the grain and provides a valuable source of energy.

Usos de *Jatropha curcas* L: Family of *Euphorbiaceae*, is a genus of herbs, bushes and distributed in tropical and subtropical trees parts of the world, mainly in África and América. Nine species have been recognized in India, some of them are grown in gardens for their ornamental foliage and flowers. However, *Jatropha curcas L.* can be used for rapid reforestation of plateaus for eco-rehabilitation and bioestéticas reasons. It can also be used as filler of all parcels in disrepair found in rural and urban areas and also to keep weeds under control unpleasant (Patil, 2007). In pasta fortified with 10% *Jatropha curcas L.* flour and semolina 90% where the protein content was higher (13.6%) compared to commercial wheat semolina pasta (12%) (Flores, 2008). In flour tortillas fortified with 10% flour *Jatropha curcas L.*, presented a greater and better nutritional value than those currently on the market

(García, 2008). In white bread roll type fortified with 5% defatted *Jatropha curcas L.* where increased protein observed (Martínez, 2010). Fortified bread box with 12.5% of protein concentrates Lupinus albus and *Jatropha curcas* which a high content of protein was obtained (López, 2010).

METHODS AND MATERIALS

Medium strong wheat (*Triticum aestivum*) flour for baking was obtained from a local milling company (Molino Elizondo, Mexico City, Mexico). Jatropha (*Jatropha curcas*) flour seeds (kindly donated by CEPROBI-IPN, Yautepec, Morelos, Mexico) and Barley protein concentrated were elaborated as follows. Jatropha flour was elaborated first dehulling the seeds with 40C water soak overnight. Dry seeds were milled using a laboratory disc mill and sieved on an 8XX sieve.

Dough Textural Properties

Dough were prepared by mixing dry ingredients: 250 g wheat flour or mix of wheat flour with Barley protein concentrate and/or Jatropha flour (according to experimental design in Table 1), plus 1 g sugar, 2 g dry yeast, 2 g salt and 12 g nonfat dry milk with water (160 mL, approximately 60% moisture) during 5 min before adding vegetable lard (Manteca Inca, Unilever, Estado of México, Mexico) and mixed until total incorporation in a Kitchen Aid 600 Stand Mixer (KitchenAid, St. Joseph, MI). After 30–32 min of fermentation at 30C, the dough was punched down to remove gases and stored in polyethylene bags under refrigeration (4 a 1°C) until analysis.

Table 1. Different formulations employed for dough and bread elaboration

Test	Wheat Flour (%)	Barley protein concentration (%)	*Jatropha curcas* Flour (%)
0	100	0	0
1	97.5	1.25	1.25
2	95	2.5	2.5
3	92.5	3.75	3.75
4	90	5	5
5	98.7	0	1.25
6	97.5	0	2.5
7	96.3	0	3.75
8	95	0	5
9	98.7	1.25	0
10	97.5	2.5	0
11	96.3	3.75	0
12	95	5	0

Dough Adhesiveness Test. Dough adhesiveness was carried out in same texture analyzer employing the SMS/Chen-Hoseney dough stickiness rig (Texture Technologies Corp.) and 2.5 mm Perspex cylinder probe (Texture Technologies Corp.)(Chen and Hoseney 1995).

Dough Extensibility Test. Dough extensibility was determined by Kieffer dough extensibility test and was carried out in same texture analyzer with the Kieffer dough and gluten extensibility rig (Texture Technologies Corp.).

Dough TPA. For TPA, 10 g of dough sample was compressed to 20% of their original height with a 2.5 cm diameter acrylic probe in a TA-XT2i texture analyzer (Texture Technologies Corp., Scarsdale, NY/Stable Micro Systems Ltd, Surrey, England) equipped with a 5 kg load cell (Szczceniak 1963).

Bread Textural Properties. Bread TPA was conducted on bread samples stacking two slices (sample thickness was 15.5 mm) from bread center, according to the recommended by Guemes et al. (2008); Totosaus, Lopez & Guemes (2013).

RESULTS AND DISCUSSIONS

Barley Concentrate and *Jatropha* Flour Addition Effect on Dough Adhesiveness

Table 1 show the dough adhesiveness results in function of adhesive force and cohesiveness. Treatments 5 to 11 presented a significantly ($p < 0.05$) difference as compared to control samples, whereas treatments 1, 2, 3, 4 and 12 had no significantly ($p > 0.05$) difference. Treatment 2 (19.7 kgf and 0.96 cm^2, adhesive force and cohesiveness, respectively) was the closer to control sample (20.1 kgf and 1.10 cm^2, adhesive force and cohesiveness, respectively). Dough adhesive force is the work necessary to overcome the attractive force between the sample and surface. Dough that presented a higher force was treatment 7 (35.4 kgf), and the lower force was treatment 2 (19.7 kgf). It can been observed that all the treatments (except treatment 2) decreased dough adhesive force, probably due to the increase in *Jatropha* flour and barley concentrate proportions, since when fiber was increased the gluten network were interrupted, reducing dough adhesiveness, according to Güemes et all. (2004, 2008). Results obtained are in concordance to those reported by Garcia (2008) with dough elaborated with wheat flour and *Jatropha* for tortillas. Control dough adhesive force (20.2 kgf) was higher than the treatment with 2.5% of *Jatropha* flour (21.0 kgf), and lower than treatment employing 10% of *Jatropha* flour (39.8 kgf). At this respect, Cruz (2008) reported that the dough adhesiveness can be interpreted as the stickiness, since changes in formulation increased dough adhesiveness since the amount of proteins and fiber allows better dough handling decreasing adhesiveness.

Barley Concentrate and Jatropha Flour Addition Effect on Dough Extensibility

Table 2 shows the barley concentrate and *Jatropha* flour addition effect on dough extensibility. For resistance, treatment 4 to 9 showed a significantly ($p < 0.05$) difference as compare to control. Treatment 7 (40.6 kgf) is the closer value to control (40.1 kgf), as well as the lower one, whereas the treatment 12 (106.3 kgf) is the higher value. The treatment 3 (71.6 kgf) is in agree to the reported by Baños (2007) for wheat flour supplemented with 5% of soluble fiber (69.4 kgf). This means that dough resistance increase when *Jatropha curcas* flour and/or barley concentrate was added, together or in different formulation. About this, Rao et all. (1992) reported an increase in resistance to extension effect by gums addition, where the extensibility decreased by the addition of gels. In other study, Palmer (1982) reported that cellulose fibers are polysaccharides, and this can affect the wheat flour physical properties, since mixed time is optimized by the presence of fiber. Park et all (1997) reported that dough made with 5% of Physillium fiber showed undesirable fragments that complicate the dough mixing, although these fragments disappeared when bread was molded-in.

For elasticity, treatments 4, 8, 9, 10 and 11 showed a significantly ($p < 0.0.5$) difference as compared to control, but samples 1 to 3, 5 to 7 and 12 were statistically equal to control. Treatment 3 (-23.4 cm) was the closer to control sample (-23.7 cm). Treatment 1 (-25.4 cm) was the higher, and treatment 10 (-15.6) was the lower one. These results are in agree to the reported by Repetsky and Klein (1981), who increased protein level in dough decreased elasticity. The increase of protein and decrease in this parameter can be observed in treatments 1 to 4 (Table 2).

For resistance work, treatments 4, 7, 9 and 10 presented no significantly ($p > 0.05$) difference as compared to control. Treatment 9 (246.3 cm^2) was the lower value and closer to control (259.3 cm^2) and treatment 2 (609.9 cm^2) was the higher one. Values are similar to those reported by Garcia (2008) studying tortillas dough with *Jatropha* flour. Minimum value reported was 274.6 cm^2 with 20% of *Jatropha* flour, and the maximum was 616.7 cm^2 with 2.5%, closer to control sample (660.8 cm^2). The increase in protein due to *Jatropha* flour resulted in a wheat protein dilution, decreasing gluten network formation resulting in more extensive dough. This also can be due to the fiber contained in barley concentrate.

Table 1. Barley concentrate and *Jatropha* flour addition effect on dough adhesiveness

Treatment	Adhesive force (kgf)	Cohesiveness (cm^2)
Control	20.1^{g}	1.1^{ef}
1	24.9^{efg}	1.7^{ef}
2	19.7^{g}	0.96^{f}
3	25.0^{efg}	1.6^{ef}
4	23.5^{fg}	1.6^{ef}
5	31.1^{bcd}	2.7^{cd}
6	31.8^{abc}	2.8^{cd}
7	35.4^{ab}	3.8^{b}
8	34.5^{abc}	5.1^{a}
9	36.6^{a}	5.3^{a}
10	29.5^{cde}	3.2^{bc}
11	32.0^{abc}	3.9^{b}
12	26.1^{def}	2.1^{de}

a, b Means with same letter in same column are not significantly different ($p > 0.05$) according to Duncan test.

Table 2. Barley concentrate and *Jatropha* flour addition effect on dough extensibility

Treatment	Resistency (Kgf)	Elasticity (cm)	Resistence work (cm^2)
Testigo	40.1[d]	-2.37[ef]	259.3[e]
1	78.1[b]	-2.54[f]	525.9[ab]
2	97.6[a]	-2.20[cde]	609.9[a]
3	71.6[b]	-2.34[ef]	483.4[bc]
4	44.2[cd]	-1.98[bcd]	301.0[de]
5	54.3[cd]	-2.27[def]	399.8[cd]
6	51.0[cd]	-2.05[bcde]	381.8[cd]
7	40.6[cd]	-2.32[ef]	335.3[de]
8	50.3[cd]	-1.74[ab]	373.3[cd]
9	42.8[cd]	-1.74[ab]	246.3[e]
10	55.4[c]	-1.56[a]	329.8[de]
11	74.0[b]	-1.93[bc]	448.6[bc]
12	106.3[a]	-2.09[cde]	545.8[ab]

[a, b] Means with same letter in same column are not significantly different ($p > 0.05$) according to Duncan test.

Barley Concentrate and Jatropha Flour Addition Effect on Dough Texture Profile Analysis

Table 3 present the behavior of the dough for hardness, adhesiveness, cohesiveness and springiness during texture profile analysis formulated with barley concentrate and *Jatropha* flour.

For dough hardness, treatments 5, 7, and 11 resulted significantly ($p < 0.05$) more hard that control. Treatment 10 (246.7 kgf) was the higher value and closer to control (279.3 kgf). This is, this dough was more easy to break and need more force to be deformed (Castillo, 2005). *Jatropha* flour and barley concentrate decreased hardness, where treatment 7 (142.3 kgf) was the lower value with 3.7% of *Jatropha* flour in the formulation. Dough become more sticky, with the unfavorable gluten network break down, reflected in troubles for its handling and processing (Hoseney, 1985). This results are in agree with the reported by Garcia (2008), in tortilla dough fortified with *Jatropha* flour, where the maximum hardness value was 258.4 kgf with 2.5% of *Jatropha* flour, and the minimum was 117.8 with 15% of *Jatropha* flour. Similar results were reported by Marquez (2007) in bakery products with different barley varieties. To employ 50% of wheat flour with 50% of barley flour obtained a hardness value of 184.0 kgf. Barley flour increased dough hardness, where 70% of barley flour was 169 kgf and 100% barley flour was 140 kgf. Totosaus (2004) reported that hardness is the force detected during the first compression cycle, indicating that the incorporation of more protein to form a more compact gel in dough requires more deformation force.

The adhesiveness of treatments 8 to 12 presented significantly ($p < 0.05$) differences as compared to control. Treatments 1 to 7 presented no significantly ($p > 0.05$) difference as compared to control (-210.0 cm^2), where sample 2 (-140.6 cm^2) was the closer to control sample. The results of dough formulated with 1.25% and 2.5% of *Jatropha* flour (treatment 5 and 6, -340.1 cm^2 and 369.2 cm^2, respectively) was not similar to the reported by Garcia

(2008), where employ 2.5% of *Jatropha* in tortillas dough resulted in -314.0 cm^2 of adhesiveness. When *Jatropha* flour was combined with barley concentrate in different proportions, like in treatments 1 to 4, adhesiveness decreased with the incorporation of these flours, with the lower value of -33.3 cm^2 in treatment 4. In treatments 5 to 12, the increase in adhesiveness due to the separate addition of *Jatropha* flour or barley concentrate doesn't present any pattern, with higher adhesiveness values in treatments 9 to 12. Fiber presence affecting hydration properties of gluten network, as reported by Güemes (1998), could be the explanation.

For cohesiveness, treatments 4 to 12 are significantly ($p < 0.05$) different to control, whereas treatments 1, 2 and 3 were not different to control sample (0.57), the closer values was for treatment 2 (0.55). This can be explained by the combination of *Jatropha* flour and barley concentrate fiber, since the cellulose contained can act like a fibrous structure to improve the physical form of the product (Crosby and Ang, 2005).

An important factor in bread elaboration is the bread elasticity, interpreted as the capacity of the sample to recover their original form or length after remove the applied force. In this view, poor dough elasticity means dough hard to handle before cut.

In springiness, there was significantly ($p < 0.05$) difference among control samples (0.72 cm) and treatment 3 (0.43 cm) and 4 (0.36 cm), being the lower values as well. Although springiness decreased with the increase in *Jatropha* flour and barley concentrate, the mixture of both flours in different proportions do not decrease dough springiness. A different behavior was observed when *Jatropha* flour or barley concentrate was added alone, where dough springiness increased as in treatment 12 (0.94 cm). Treatments 5 to 8, elaborated with *Jatropha* flour, obtained the higher (0.93 cm) value in treatment 7, indicating that springiness was increased by the protein contained in *Jatropha* flour in separate way to fiber content in barley concentrate.

Table 3. Barley concentrate and *Jatropha* flour addition effect on dough texture profile analysis

Treatment	Hardness (kgf)	Adhesiveness (cm^2)	Cohesiveness	Springiness (cm)
Control	279.3^{a}	-210.0^{ab}	0.57^{bc}	0.72^{a}
1	207.9^{abc}	-107.9^{a}	0.53^{c}	0.66^{ab}
2	216.8^{abc}	-140.6^{a}	0.55^{c}	0.66^{ab}
3	241.9^{ab}	-94.8^{a}	0.45^{cd}	0.43^{bc}
4	225.5^{abc}	-33.3^{a}	0.39^{d}	0.36^{c}
5	168.7^{bc}	-340.1^{bc}	0.74^{a}	0.91^{a}
6	214.1^{abc}	-369.2^{bcd}	0.69^{ab}	0.81^{a}
7	142.3^{c}	-423.2^{cde}	0.83^{a}	0.93^{a}
8	207.1^{abc}	-554.4^{e}	0.81^{a}	0.92^{a}
9	223.2^{abc}	-537.3^{de}	0.78^{a}	0.93^{a}
10	247.6^{ab}	-565.5^{e}	0.77^{a}	0.93^{a}
11	192.5^{bc}	-464.3^{cde}	0.75^{a}	0.92^{a}
12	210.3^{abc}	-504.6^{cde}	0.74^{a}	0.94^{a}

a, b Means with same letter in same column are not significantly different ($p > 0.05$) according to Duncan test.

Table 4. Barley concentrate and *Jatropha* flour addition effect on bread texture profile analysis

Treatment	Hardness (kgf)	Cohesiveness	Springiness (cm)
Testigo	1.52[b]	0.40[b]	0.71[a]
1	2.52[a]	0.48[ab]	0.74[a]
2	2.00[ab]	0.50[ab]	0.72[a]
3	1.51[b]	0.57[a]	0.95[a]
7	1.43[b]	0.60[a]	0.92[a]

[a, b] Means with same letter in same column are not significantly different ($p > 0.05$) according to Duncan test.

Barley Concentrate and *Jatropha* Flour Addition Effect on Bread Texture Profile Analysis

Table 4 presents the textural profile analysis of bread elaborated with *Jatropha* flour and barley concentrate. For bread hardness, significantly ($p < 0.05$) higher values were observed in treatment 1 (2.52 kgf) as compared to control (1.52 kgf), whereas treatments 2, 3, and 7 was not different. Bread of treatment 3 (1.51 kgf) was the closer to control, both with lower values. For bread cohesiveness, treatments 3 and 7 were significantly ($p < 0.05$) different to control, and treatments 1 and 2 were not different. Park et all. (1997) reported that bread firmness was improved in additive-less bread when 5% of fiber was employed, showing minimum firmness but acceptable loaf volume. Czuchajowska et all. (1992) previously reported that to use 4% of fiber increased bread softness. For springiness, there was no significantly ($p > 0.05$) difference among treatments.

CONCLUSION

Dough adhesive force and cohesiveness presented no changes as compared to control when both flours were incorporated.

Resistance and resistance work of dough was higher with the increase of Jatropha flour or barley concentrate.

Texture profile analysis of dough showed that adhesiveness, cohesiveness and springiness decreased when both flours were employed separately, with an inverse effect on dough hardness.

In bread texture profile analysis cohesiveness and springiness decreased as compared to control, with similar hardness values when 3.75% of each flour was employed.

REFERENCES

Baños, B. (2007). Caracterización Reológica de Masas de Harina de Trigo Adicionadas con Fibra Soluble. Tesis de Licenciatura. Universidad Autónoma del Estado de Hidalgo. Instituto de Ciencias Agropecuarias. México, D.F.

Belitz, H. D. & y Grosch, W. (2002). *Food Chemistry*. 3da Edición. Ed. Springer. Zaragoza, España. 696-707.

Calaveras, J. (2004). Tratado de Panificación y Bollería. AMV Ediciones. Madrid, *España*. 187-202.

Callejo, G. M. J. (2002). Industrias de Cereales y Derivados. Ediciones Mundi-Prensa. Madrid. 24-35, 62-67, 90-101, 191-208, 222-232.

Castillo, R. C. (2005). Uso Alternativo de las Cebadas (*Hordeum sativum jess*) Cultivadas en el Estado de Hidalgo y Tlaxcala en la Industria de Panificación. Tesis de Licenciatura. LQA-UAEH.

Chen, W. Z. & Hoseney, R. C. (2002). Wheat Flour Compound that Produces Sticky Dough: Isolation and Identification, *J. Food Sci.*, *60*(3), 434-437.

Crosby G. A. & y Ang J. F. (2005). Formulating reduced-calorie foods with powdered cellulose. A publication of the Institute of Food Technologists. *Food Tech. March*, Vol. *59*, No. 3, 35-38.

Cruz-Victoria, M. T., Contreras, K. E. & Anaya S. I. (2008). Aceite de la *Jatropha curcas*, análisis de su composición. Escuela Nacional de Ciencias Biológicas. Instituto Politécnico Nacional.

Czuchajowska, Z., Szczodrak, J. & Pomeranz, Y. (1992). Characterization and Estimation of Barley Polysaccharides by Near-Infrared Spectroscopy. I. Barleys, Starches and B-d-glucans. *Cereal Chem.*, *69*, 413-418.

Desrosier, N. (1999). Elementos de Tecnología de Alimentos. Editorial CECSA, S. A.

Flores, G. (2008). Evaluación Química, Reológica y Sensorial de Pastas Fortificadas con Harina de *Jatropha curcas*. Tesis de Licenciatura. Universidad Autónoma del Estado de Hidalgo. Instituto de Ciencias Agropecuarias. México, D.F.

García, I. (2008). Fortificación de de Tortillas de Harina de Trigo con Piñón Silvestre (*Jatropha curcas* L.). Tesis de Licenciatura. Universidad Autónoma del Estado de Hidalgo. Instituto de Ciencias Agropecuarias. México, D.F.

Granfeldt, Y., Liljeberg, H., Drews, A., Newman, R. & Bjorck, I. (1994). Glucose and Insuline Responses to Barley Products: Influence of Food Structure and Amylose-Amylopectin ratio. *Am. J. Clin. Nutr.*, *59*, 1075-1081.

Güemes, N. (1998). Efecto de la Adición de Sustitutos de Grasa a Base de Carbohidratos Sobre las Características de Donas Fermentadas y no Fermentadas. Tesis de Maestría. Instituto Politécnico Nacional. Escuela Nacional de Ciencias Biológicas. México, D.F.

Güemes-Vera, N., Arciniega-Ruiz Esperanza, O. & Dávila-Ortiz, G. (2004). Structural Análisis of the *Lupinus mutabilis* Seed, its Flour, Concentrate, and Isolate as Well as Their Behavior When Mixed with Wheat Flour. *Lebensmittel-wissenschaft Und-Technlogie*. *37*, 3, 283-290.

Güemes-Vera, N., Peña-Bautista, R., Jiménez-Martínez, C., Dávila-Ortiz, G & Calderón-Domínguez, G. (2008). Effective Detoxification and Decoloration of *Lupinus mutabilis*

Seed Derivatives on Bread Quality and Acceptance. *Journal of the Science of Food and Agriculture. 88*, 1135.1143.

Hoseney, C. R. (1985). Principios de Ciencia y Tecnología de los Cereales. Ed. Acribia, Zaragoza, España.

Kent N. L. Tecnología de los Cereales, Introducción para Estudiantes de Ciencias de los Alimentos y Agricultura. Ed. Acribia. 1era. ed. Zaragoza, España. 1987. 133-142.

López, H. (2010). Propiedades Funcionales y de Textura de Masa y Pan de Caja Fortificados con Concentrados Proteicos de *Lupinus (Lupinus albus) y Jatropha (Jatropha curcas)*. Tesis de Maestría. Tecnológico de estudios superiores de Ecatepec. México, D.F.

Márquez, A. (2007). Elaboración y Evaluación de un Producto de Panificación Elaborado con Harina de Cebada. Tesis de Licenciatura. Universidad Autónoma del Estado de Hidalgo. Instituto de Ciencias Básicas e Ingeniería. México, D.F.

Martínez, H. J. (2004). The Nutritional Potential of Mexican piñon (*Jatropha curcas*). Toxic and Antinutritional Factors. Centro de Desarrollo de Productos Bióticos. Instituto Politécnico Nacional. Yautepec, Morelos, México.

Martínez, H. J. (2006). *Jatropha curcas*, Cultivating Economic Growth Toxic and antinutritional factors. Centro de Desarrollo de Productos Bióticos. Instituto Politécnico Nacional. Yautepec, Morelos, México.

Martínez, H. J. (2007). Experiencia con *Jatropha curcas* L. en México. Centro de Desarrollo de Productos Bióticos. Instituto Politécnico Nacional. Yautepec, Morelos, México.

Martínez, G. (2010). Pan Blanco Tipo Bolillo Fortificado con Harina de *Jatropha curcas*. Tesis de Licenciatura. Universidad Autónoma del Estado de Hidalgo. Instituto de Ciencias Agropecuarias. México, D.F. En Revisión.

Newman. R. K. & Newman, C. W. (1991). Barley as a Food Grain. *Cereal Foods World, 36*, 800-805.

Palmer, J. K. (1982). Carbohydrate in Sweet Potato. In Sweet Potato. Proceedings of First International Symposim, R.I. Villareal and T.D. Griggs (Ed.), p.135. Asian Vegetables Research and Development center, Taiwan.

Park. P. A. Seib & Chung, O. K. (1997). Dietary and Composition in Home-Prepared and Commercially Baked Products: Analysis and Prediction. *Cereal Chemistry., 71*(1), 99-103

Patil, V. R. (2007). *Jatropha* – an alternative to diesel. *Agricultura and Industry Survey*, Vol. *13*, Issue No. 5, Vadamalai Media Group, Bangalore.

Rao, M. A. (1992) Rheological Properties of Fluid Foods en Engineering Properties of Foods, Ed. Marcel Dekker, New York.

Repestsky, J. A. & Klein, B. P. (1981). Partial Replacement of Wheat Flour with Yellow Field Pea Flour in White Bread. *Journal of Food Science*. Vol *47*. pp 327.

Repestsky , J. A. & Klein, B. P. (1981). Partial Replacement of Wheat Flour with Yellow Field Pea Flour in White Bread. *Journal of Food Science*. Vol *47*. pp 327.

SAS/STAT. 2000. Users Guide Release 603. Edition Cory N.C. USA.

Schneeman, B. O. (1999). Fiber, Inulin and Oligofructose: Similarities and Differences. *J Nutr.*, 129, 1424S-1427S.

Serna-Saldívar, S. R. O. (2001). Química, Almacenamiento e Industrialización de los Cereales. AGT Editor, México, D.F. 3-23, 47-73 y 79-89.

Szczodrak, J., Czuchajowsca, Z., & Pomeranz, Y. (1992). Characterization and Estimation of Barley Polysaccharides by Near-infrared Spectroscopy. II. Estimation on Total B-D-glucans. *Cereal Chem.*, *69*, 419-423.

Texture Technologies (2002). The Texture Report. A newsletter for the Academic Clients of Texture Technology Corp.

Totosaus, A. (2004). Funcionality of Glicosilated Heart Surimi and Heat-Precipitated Whey Proteins in Meat Batters. *J. Muscle Foods*. V. *15*, 256-268.

Totosaus A. Lòpez H & Guemes-Vera, N. (2013). Effect of Lupinus (*Lupinus Albus*) and Jatropha (*Jatropha Curcas*) Protein Concentrates on Wheat Dough Texture and Bread Quality: Optimization by A D-Optimal Mixture Design. *J. Texture Studies*, *44*(4), 1-12.

INDEX

B

C

D

E

F

G

H

M

N

O

P

Q

R

S

T

U

V

W

Y

Z